Hettner-Lectures, 6

Geography, gender, and the workaday world

HETTNER-LECTURES

Series Editors: Hans Gebhardt and Peter Meusburger

Managing Editor: Michael Hoyler

Hettner-Lectures, 6

Department of Geography, University of Heidelberg
in association with

Franz Steiner Verlag Stuttgart

2003

GEOGRAPHY, GENDER, AND THE WORKADAY WORLD

Hettner-Lecture 2002

with

Susan Hanson

School of Geography
Clark University
Worcester, MA
USA

Department of Geography, University of Heidelberg
in association with

Franz Steiner Verlag Stuttgart

2003

First published 2003

Department of Geography
University of Heidelberg
Berliner Straße 48
D-69120 Heidelberg
Germany

http://www.geog.uni-heidelberg.de/

Cover illustration: Manufacture of globes, France c.1925 © Archiv für Kunst und Geschichte, Berlin (akg-images); photograph of Susan Hanson by Pia Volk
Cover design: Michael Hoyler and Christine Brückner

Illustrations in 'Geography, gender, and the workaday world': Figure 1 (p. 7) from the Internet; Figure 2 (p. 8) reprinted from *Animal Behavior* 49 (1995), Deborah Gordon, 'The development of an ant colony's foraging range', pp. 649-59, with permission from Elsevier; Figure 3 (p. 9) reprinted from *Journal of Animal Ecology* 63 (1994), Paul M. Thompson, David Miller, Richard Cooper, and Philip S. Hammond, 'Changes in the distribution and activity of female harbour seals during the breeding season: implications for their lactation strategy and mating patterns', pp. 24-30, with permission from Blackwell; Figure 4 (p. 10) by Susan Hanson; Figure 5 (p. 10) from the Internet; Figure 6 (p. 11) from Walter Christaller, *Central places in Southern Germany* (Englewood Cliffs, N.J.: Prentice Hall, 1966); Figure 7 (p. 14) reprinted from Susan Hanson and Geraldine Pratt, *Gender, work, and space* (London: Routledge, 1995), with permission from Routledge; Figures 8 (p. 17) and 9 (p. 18) reprinted with permission from *Urban Geography* 9 (1988), No. 2, Susan Hanson and Geraldine Pratt, 'Spatial dimensions of the gender division of labor in a local labor market', pp. 180-202. © V.H. Winston & Son, Inc., 360 South Ocean Boulevard, Palm Beach, FL 33480; Figure 10 (p. 19) reprinted from Susan Hanson and Geraldine Pratt, *Gender, work, and space* (London: Routledge, 1995), with permission from Routledge; Figure 11 from National Personal Travel Survey 1995; Figure 12 by Susan Hanson. All rights reserved.

All photographs in 'Photographic representations' by Pia Volk

Printed in Germany by Druckagentur Jürgen-J. Sause, Heidelberg

ISBN 3-515-08369-3

Contents

INTRODUCTION

Introduction: Hettner-Lecture 2002 in Heidelberg

PETER MEUSBURGER and HANS GEBHARDT

The Department of Geography, University of Heidelberg, held its sixth 'Hettner-Lecture' from July 1-5, 2002. This annual lecture series, named after Alfred Hettner, Professor of Geography in Heidelberg from 1899 to 1928 and one of the most reputable German geographers of his day, is devoted to new theoretical developments in the crossover fields of geography, economics, the social sciences, and the humanities.

During their stay, the invited guest-speakers present two public lectures, one of which is transmitted via teleteaching on the Internet. In addition, several seminars give graduate students and young researchers the opportunity to meet and converse with an internationally acclaimed scholar. Such an experience at an early stage in the academic career opens up new perspectives for research and encourages critical reflections on current theoretical debates and geographical practice.

The sixth Hettner-Lecture was given by Susan Hanson, Professor of Geography at Clark University, Worcester, MA. Susan Hanson is widely known for her research in urban, economic, and feminist geography, focusing on issues of urban transportation (e.g. *The geography of urban transportation* (2nd ed. 1995)), gender and urban labour markets (e.g. *Gender, work, and space* (with Geraldine Pratt, 1995)), and sustainable practices in urban areas. Susan Hanson is a past president of the Association of American Geographers and was the first woman geographer to be elected to the National Academy of Sciences in 2000. She is also a member of the American Academy of Arts and Sciences and received the American Geographic Society's Van Cleef Medal for outstanding work in urban geography in 1999.

During the Hettner-Lecture 2002 Susan Hanson presented two public lectures entitled 'Gender, geography, and the workaday world' and 'Entrepreneurship: geographical and feminist perspectives',[1] both of which are published here in revised form. Three seminars with graduate students and young researchers from Heidelberg and nine other European universities took up issues raised in the lectures. The seminars were entitled 'What the !#$&* is feminist geography?', 'Contextualizing entrepreneurship', and 'Combining quantitative and qualitative methods in human geography'.

---

[1] 'Gender, geography, and the workaday world', *Alte Aula der Universität*, Monday, 1st July 2002, 18.15; afterwards reception, *Bel Etage, Rector's Office*. 'Entrepreneurship: geographical and feminist perspectives', *Hörsaal des Geographischen Instituts*, Tuesday, 2nd July 2002, 15.15.

We should like to express our gratitude to the *Klaus Tschira Foundation* for generously supporting the Hettner-Lecture. Particular thanks are due to Dr. h.c. Klaus Tschira for his continuing interest in frontiers of geographical research.

The Hettner-Lecture 2002 would not have been possible without the full commitment of all involved students and faculty members. Once again, Tim Freytag, Michael Hoyler and Heike Jöns were crucial in all organisational and conceptual matters. We are also grateful to the students who helped with the organisation of the event. The concerted effort and enthusiasm of all participants once more ensured a successful Hettner-Lecture in Heidelberg.

GEOGRAPHY, GENDER, AND THE WORKADAY WORLD

Geography, gender, and the workaday world *

SUSAN HANSON

This lecture tonight is a tale of research adventures. It is about the inchworm, with one end anchored in where it has been and the other end exploring new directions.[1] It is a tale about the melding of people and place, of agency and structure, of spatial analysis and social theory, of geography and feminism. But most of all it is a tale about the enduring power of the geography of everyday life.

Figure 1 I include this cartoon of an inchworm after lengthy discussions in local beer halls revealed that this creature does not translate easily into German

I am the inchworm (Figure 1). For more than 30 years I've been inching along, trying to understand various aspects of urban life, not from some inchworm's version of a large and cozy armchair in which I might sit and ponder the nature of cities, but from going out and collecting and analyzing data, engaging with the real world on the ground, very much like an inchworm. I've been involved in three large studies, each of which collected and analyzed a lot of data on people in cities. In each case, even though I started out with the idea that I was studying something else – something that might be big and important and new and exciting – after a while I had to face the fact that what I was really studying was the mundane geography of plain old ordinary everyday life. Over the years I have come to appreciate the power of the insights to be gained through an understanding of the geography of everyday life, the leverage

---

* My thanks to Peter and Anne-Marie Meusburger for their warm hospitality in Heidelberg. I am grateful to Peter for his suggestion that I use the Hettner lecture as an opportunity to reflect on my life as a geographer. I thank Dr Klaus Tschira, the geography faculty at the University of Heidelberg, and the many students whose feisty arguments brought enthusiasm and vitality to the seminars; special thanks to those whose thought and planning gave shape to the seminars: Tim Freytag, Michael Hoyler, and Heike Jöns.

[1] This is what academics like to call "path dependence."

that such understanding provides for developing theories about various aspects of urban life, such as how urban labor markets work.

So, I want to talk about geography, gender, and everyday life – the workaday world. I hope that this focus on the mundane topic of everyday life will prompt you to think how it might relate to your own work on problems that at first glance might seem not to be related to everyday life. My talk tonight has four main parts. First I'll explain what I mean by the geography of everyday life. Second, I'll briefly describe each of the three studies that I've helped to design and lead. Third, I'll go into considerable detail about one of the studies, the one that focused on gender and urban labor markets, as an example of how the geography of everyday life lies beneath the processes and structures that shape cities. Finally, I'll conclude with some thoughts about why understanding geography, gender, and the workaday world is so important and central to our work as geographers and social scientists.

The geography of everyday life

16 July 1991

19 July 1991

1991 Cumulative

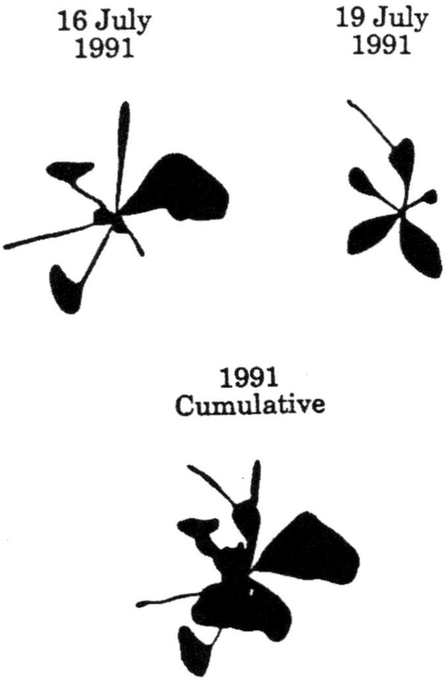

Figure 2 Foraging maps of a colony of ants for two separate days and cumulatively over many days

First, what do I mean by the geography of everyday life? Let's begin with ants (Figure 2). These maps of the foraging spaces of an ant colony show where the everyday life of the members of this ant colony in the Arizona desert was carried out, for each of two separate days and then for several days in 1991. We can move from the everyday lives of ants to the everyday lives of people by way of the everyday lives of seals (Figure 3). This map shows the foraging space of an adult female seal off the coast of Scotland over the course of a breeding season.

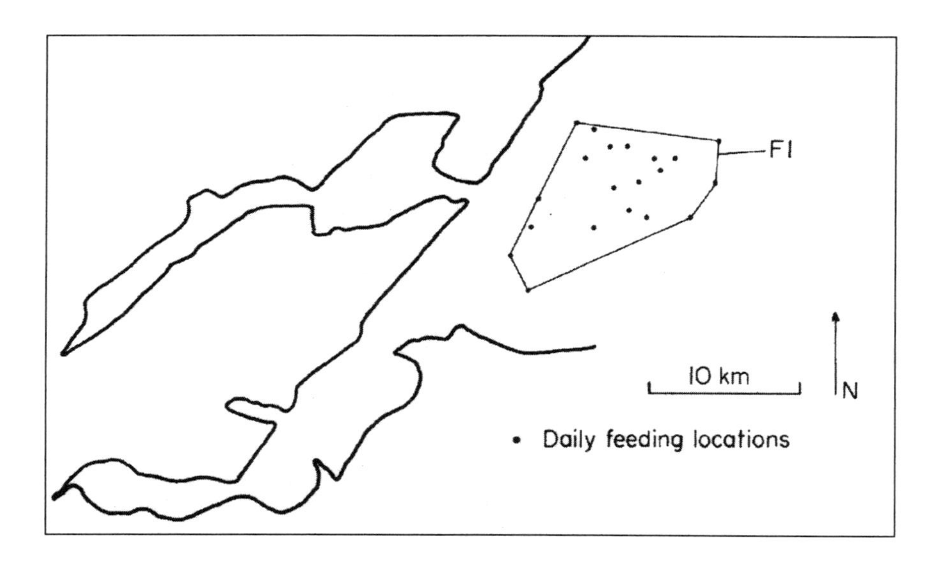

Figure 3 Feeding space of one adult female harbor seal off the coast of Scotland

By the geography of everyday life – whether of ants or seals or humans – I mean the daily activity patterns and activity spaces that map out daily routines, like those shown here for a hypothetical person (Figure 4). In the case of humans, the "geography of everyday life" refers to what people do (that is, what activities they engage in), where they do these activities, how they get there (that is, what means of travel they use, what routes they follow), and with whom they interact. Someone has even suggested that we could think of such activity patterns as defining what constitutes the city for each person or household; that is, each person or household creates its own city in the set of places it incorporates into its activity pattern.[2]

---

[2] Robert Fishman 1990, 'America's new city',. *The Wilson Quarterly* 14 (1990) pp. 25-45.

Three universal principles underlie the geography of everyday life: (1) everyone – rich and poor, man and woman, young and old – has 24 hours in a day; (2) no one can be in more than one place at a time, and (3) no one can move instantaneously from one place to another, that is, we have no truly magic carpets yet (Figure 5).[3] People are bound by time and space.

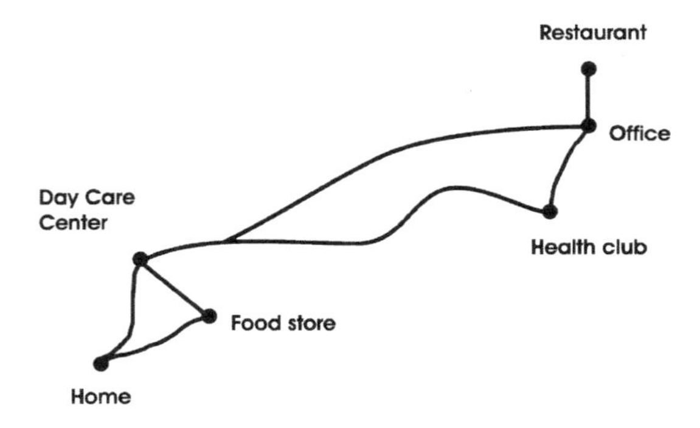

Figure 4 Sketch map of one daily human activity pattern; the foraging space of one adult human

Figure 5 Until technology brings us magic carpets, we are bound by space and time

Now, you may say that I'm presenting a very old-fashioned view of everyday life; what about cell phones? What about the Internet? What about video conferencing? What about the death of distance? Technology may indeed be eroding the power of

---

[3] Susan Hanson and Perry Hanson, 'The geography of everyday life', in Reginald G. Golledge and Tommy Gärling (eds.) *Behavior and environment: psychological and geographical approaches* (Amsterdam: North-Holland, 1993) pp. 249-69.

these three principles to shape the geography of everyday life (and later I will talk a bit about the impact of technology). But the friction of distance – which in geography refers to the idea that it is more difficult and costly to interact over longer distances than over shorter ones – and the myriad processes that are caught in the friction of distance remain surprisingly durable despite distance-defying technologies.

The roots of interest in studying the geography of everyday life go deep into the history of our discipline. I'm not going to go farther back than Walter Christaller although you can no doubt think of examples that pre-date that famous German geographer. Christaller, as you know, focused on the daily travels of ordinary people to purchase goods and services; he was in a way concerned with the geography of everyday life (although as you also know, he wasn't dealing with real people, and he did do his thinking in an armchair). His creative insight was to see that through their everyday micro decisions about where to go to buy goods and services, people were creating a settlement system, a system of central places, as shown in this iconic diagram (Figure 6). He identified the link between the geography of individuals' everyday lives and urban spatial structure.

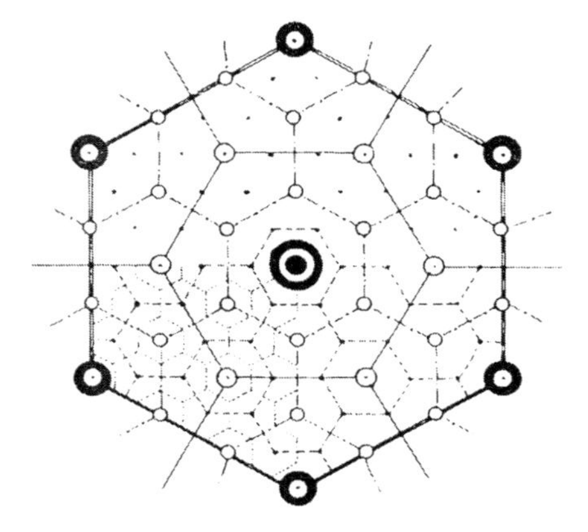

Figure 6 Christaller's hexagons

Shortly after Christaller's work became available in English (1966)[4] there was a tremendous surge of interest throughout the social sciences in what I would call the geography of everyday life – although it wasn't called that at the time. During the late 1960's, for example, scholars from many disciplines collaborated in a monumental international comparative study of time use – how people in different countries

[4] Walter Christaller, *Central places in Southern Germany* (Englewood Cliffs, N.J.: Prentice Hall, 1966) [translation by C.W. Baskin of Christaller's 1933 book].

allocate the 24-hour day to different activities.[5] Some of these scholars even looked at gender differences in time use. In urban planning, F. Stuart Chapin pointed to the importance of studying people's daily activity patterns as a way of understanding how they make use of the urban spaces and places that planners help to create.[6] In geography, Torsten Hägerstrand followed his own pioneering work on the importance of face-to-face contact in the diffusion of innovation (which of course emphasized the importance of everyday interactions) with studies of time geography and the concept of the time-space prism;[7] also in geography, studies of urban travel behavior flourished, and the concept of action space or activity space emerged.[8] Independently, it seems, in many different quarters, interest in everyday life – and the geography of everyday life – bubbled up.

How do social scientists – and especially geographers – go about studying the geography of everyday life? The map of ant foraging spaces that I showed you is the result of people observing the ant colony and marking out where the ants went over the course of a day.[9] The researchers tracking the seals used VHF radiotelemetry, which involves capturing the seals and attaching radio-tags to them.[10] Reactions to human surveillance systems already in use, such as hidden cameras, suggest that most people would object to having their activities tracked by wearing a radio tag. So, how do we study the everyday lives of human beings?

Imagine that a video camera follows you incessantly (with the tape rolling) for a day, a week, a month, a year – this is how "reality-based" TV shows supposedly capture the reality of everyday life. This video record might approximate a complete picture of your activity pattern, but it might be somewhat difficult to implement as a research protocol. Now imagine that a still camera follows you incessantly, taking snapshots at regular or random intervals, recording in each snapshot, the time, place, activity, and so on. Data collected on the geography of everyday life resemble snapshots from the still camera in that these data are necessarily selective; they focus

---

5 Alexander Szalai (ed.) *The use of time* (The Hague: Mouton, 1971).

6 F. Stuart Chapin Jr., *Human activity patterns in the city: things people do in time and in space* (New York: Wiley, 1974).

7 Torsten Hägerstrand, 'The impact of social organization and environment upon the time-use of individuals and households', *Plan, International, Special Issue* (1972) pp. 24-30; Bo Lenntorp, *Paths in space-time environments: a time-geographic study of movement possibilities of individuals* (Lund: CWK Gleerup, 1976).

8 Frank Horton and David Reynolds, 'Effects of urban spatial structure on individual behavior', *Economic Geography* 47 (1971) pp. 36-48.

9 Deborah Gordon, 'The development of an ant colony's foraging range', *Animal Behavior* 49 (1995) pp. 649-59.

10 Paul M. Thompson, David Miller, Richard Cooper, and Philip S. Hammond, 'Changes in the distribution and activity of female harbour seals during the breeding season: implications for their lactation strategy and mating patterns', *Journal of Animal Ecology* 63 (1994) pp. 24-30.

only on certain activities, to the neglect of others, but they usually do this for a large sample of people over a relatively short span of time (such as one day or one week).

Three studies of people in cities

As I mentioned, I've been involved in three large studies over the past 30 years, and I want to describe each one briefly so you can appreciate the common threads and the differences among them. After providing the overview of all three, I'll describe one in greater detail. None of these studies had the explicit goal at the outset to understand the geography of everyday life, but in each case it was the geography of everyday life that turned out to be central to the problem at hand.

The first study, in 1971, was the Uppsala Household Travel Survey, which I undertook with Duane Marble and Perry Hanson in Uppsala, Sweden. This was a study of people's travel activity patterns in the city, and it formed the basis of my doctoral thesis on travel behavior and destination choice. The core of the data we collected was the travel diary. For five weeks the adult members of 300 households kept detailed records of all of their out-of-home movements, including – for each segment of each trip – where they went (defined as a street address), how they traveled there (bus, bike, car, feet), what activity they did there (bought a book, watched a film, dropped a child off at day care), who accompanied them on the trip, and the time of arrival at and departure from each place they visited. In those pre-GIS times, we also created and digitized a complete inventory of all establishments (e.g., stores, banks, post offices, parks, factories) in the urbanized area of Uppsala.[11]

You might think that this was explicitly a study of the geography of everyday life, but that is not how we thought of it at the time. Everyday life was too trivial and mundane to be the subject of serious study. The goals that lay behind the collection of this massive database had to do with improving the predictive ability of travel-demand models for transportation planning and making generalizations about the relationships between people's travel behavior and the nature of the urban environment. Only after I became immersed in feminist thinking, with its clear emphasis on the importance of everyday life, was I able to talk about this study in terms of "the geography of everyday life."

---

[11] Kay Axhausen and colleagues recently replicated the Uppsala study in two German cities, Karlsruhe and Halle/Salle; Kay W. Axhausen, Andrea Zimmermann, Stefan Schönfelder, Guido Rindsfüser, Thomas Haupt, 'Observing the rhythms of daily life: a six-week travel diary', *Transportation* 29 (2002) pp. 95-124.

The second study is known as the Worcester Expedition,[12] which Geraldine Pratt and I – along with many students – carried out in Worcester, Massachusetts from 1987 to 1989. Because I'll be talking about this study in some detail, let me take a minute to give you a sense of Worcester as a place. Located about 50 miles west of Boston, the Worcester metropolitan area had a population of about 436,000 in 1990 (511,000 in 2000) (Figure 7). Worcester has always had a diversified economic base, but in the past 40 years the importance of manufacturing has declined (from 46% of all employment in 1960 to 21 % in 1990) while the importance of services has increased (accounting for only 4% of employment in 1960, but fully 27% in 1990). The manufacturing industries here range from Worcester's traditional strengths in wire, metal products and machine tools, abrasives, and envelopes to computer hardware and software, plastics, optics, and medical instruments. Among Worcester's important producer services are higher education, health care, and insurance. There are some 23,000 university students studying in 12 colleges and universities in the city; relative to the size of Worcester's population, the city has more college students than Boston.

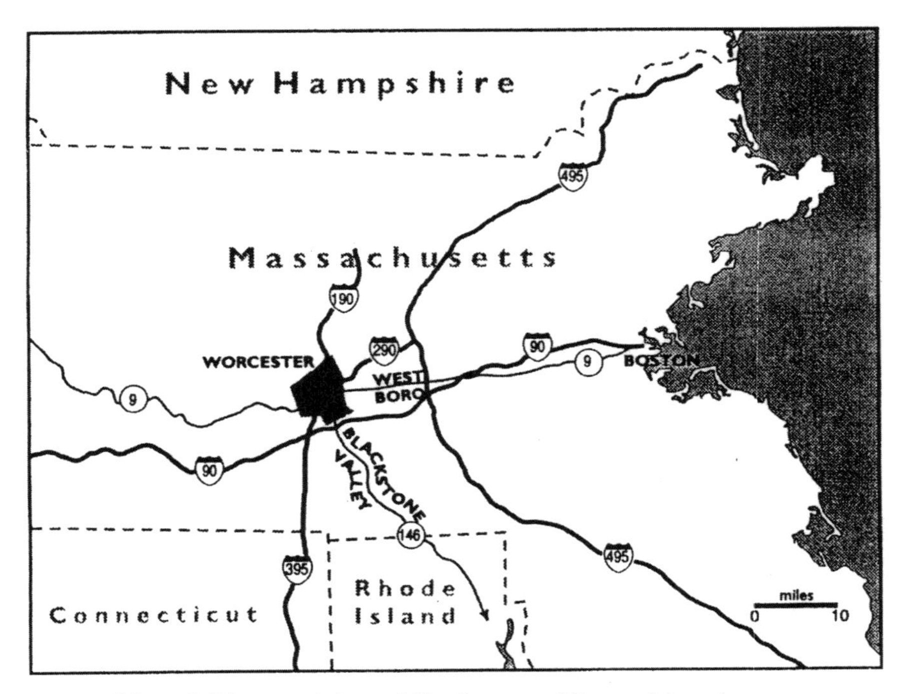

Figure 7 Worcester is located 50 miles west of Boston, Massachusetts

[12] The students who participated in the study gave it this name, in part because much of the study's funding came from the National Geographic Society, which had funded many explorations and expeditions, but never before one that explored an urban area.

The Worcester expedition was about gender and local labor markets in cities. It grew out of one very small and unexpected finding that emerged from analysis of travel diary data from Baltimore, Maryland, namely that people who work in female-dominated occupations (that is, occupations in which more than 70% of the workers are women) work closer to home than do people employed in other lines of work.[13] This finding posed the question that was the initial motivation for undertaking the Worcester study: how might geography contribute to the creation and persistence of gender-based segmentation in the labor market? Segmentation refers to the divisions that are so evident in patterns of work around the world, namely that women and men work in different industries, different occupations, and different jobs – with the result that women earn less money than men and have fewer opportunities for growth and advancement on the job.

The Worcester Expedition involved interviewing women and men in 640 households to learn about how they decide where to live and where to work, what they value in a job and in a residential setting, what kinds of work they had done and where they had lived over the past ten years, how they had found their current job and housing, and how they manage family and work responsibilities within the household. The Worcester study also involved interviewing 130 employers to learn about how they make location decisions, what they look for in the workers they hire, and how they design the labor process around the characteristics of particular labor forces.[14]

The third study is the one I'm still working on (and will be talking about in more detail during this week); it extends the labor market study to look at entrepreneurship and self-employment. With the help of many students, I have conducted in-depth interviews with about 200 business owners in each of two American cities – Worcester, Massachusetts and Colorado Springs, Colorado – to learn about how they came to start their own businesses (the nature of the start-up process), how business ownership relates to owners' previous work experiences, how business owners make location decisions, how support networks – both personal and professional – play a role, and how business owners think about their relationship to place. In each city, we also sent out a mailed survey, after the interviews, to learn about these same issues from a larger sample of business owners.

Now let's look in more detail at the labor market study to see how everyday life ambushed us in our attempt to understand the role of geography in creating and sustaining gender-based labor market segmentation.

---

[13] Susan Hanson and Ibipo Johnston, 'Gender differences in worktrip length: explanations and implications', *Urban Geography* 6 (1985) pp. 193-219.

[14] The methodology of the Worcester Expedition is described in detail in Susan Hanson and Geraldine Pratt, *Gender, work, and space* (New York: Routledge, 1995) chapter 3 and appendix.

Geography, gender, and work[15]

When we interviewed women and men in Worcester in the late 1980s, more than half (54%) of women and nearly two-thirds of men (65%) worked in gender-typical occupations, defined as occupations in which more than 70% of the workers were of their own sex. So, for example, women were over-represented in occupations such as cashier, textile machine operator, semi-conductor assembler, and kitchen worker, and men were disproportionately represented in occupations like janitor, truck driver, telephone line installer, lawyer, or physician. Similar patterns of gender-based segmentation in the labor market are found in Germany, with women over-represented (relative to their overall proportion of the workforce) in clerical and sales and service-related work and under-represented in administrative and managerial and craftsman and production work.[16]

Gender-based occupational segregation is responsible for a substantial portion of the gender wage gap, the fact that women who work full time, year round earn on average significantly less money than men earn; in the U.S. women earn an average of 74 cents for every dollar men earn, and for women of color this figure falls to 64 cents on the dollar.[17] In our Worcester sample women's hourly earnings were on average only 60% of men's ($9 per hour for women, compared to $15 for men); among women, moreover, those who worked in female-dominated occupations earned about three-quarters of what women who worked in male-dominated jobs earned ($8.70 per hour for women in female-dominated vs. $11.70 for women in male-dominated occupations). Female-dominated jobs are also less likely to have health insurance and retirement benefits[18] or flexible working hours. Germany, too, has a substantial gender wage gap, with women earning on average about 75% of what men earn.[19]

So, the segregation of women and men in the labor market is certainly visible; it's widespread (it's not just an American phenomenon); it's durable; and it has real consequences for people in terms of labor market outcomes (wages, benefits, opportunities for advancement).

---

[15] This section summarizes work reported in many publications that Geraldine Pratt and I have written together (Pratt and Hanson; Hanson and Pratt); all of these articles are listed in the reference list at the end of Hanson and Pratt, Gender, work, and space *op. cit.*

[16] Ministry of Manpower, Manpower Research and Statistics Department, Singapore, *Occupation segregation: a gender perspective*, Paper 1/00, April 2000 (*http://www.mom.gov.sg/mrsd/publication*)

[17] Heidi Hartmann, Katherine Allen, and Christine Owens, *Equal pay for working families: national and state data on the pay gap and its costs* (Washington, DC: AFL-CIO Working Women's Department, 1999).

[18] Hanson and Pratt, Gender, work, and space *op. cit.*, p. 63.

[19] ILO (International Labor Organization), *Yearbook of labor statistics* (1997).

How does geography enter into this picture? Let's start with a map. I've described how labor markets are distinctly gendered, but is there a geography to the gender division of labor? Within the Worcester metropolitan area the answer is decidedly yes (Figure 8).

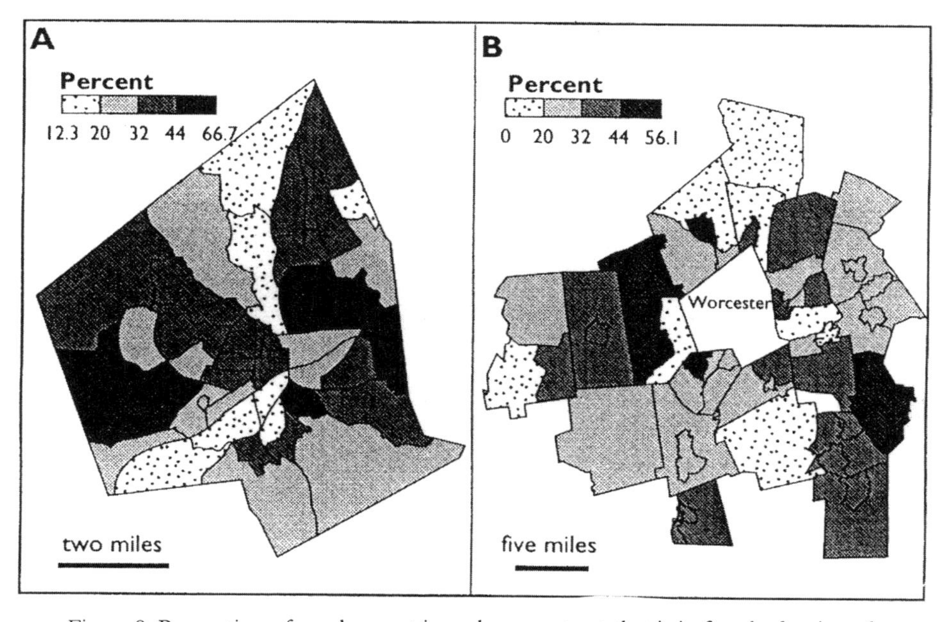

Figure 8 Proportion of employment in each census tract that is in female-dominated occupations for (A) the city of Worcester, and (B) the rest of the metropolitan area

This map shows the proportion of jobs in each census tract that are in female-dominated occupations (like clerical work, semiconductor assembler, or textile sewing machine operator); in the areas that are darker on this map a higher proportion of employment is in female-typed jobs. We created the same kind of maps for jobs in male-dominated occupations, which include jobs like fork lift operator or machinist (Figure 9). These maps show that different job types (i.e., women's jobs vs. men's jobs) – and different pay scales and different norms and labor cultures – are found in different parts of the metro area. We found similar patterns when we controlled for industry type, so that, for example, if we map employment within the industry category health, education, and welfare or within manufacturing by gender typing of job, we find the same result: jobs held by women in female-dominated occupations are located in different parts of the city than are jobs held by men in male-dominated occupations.

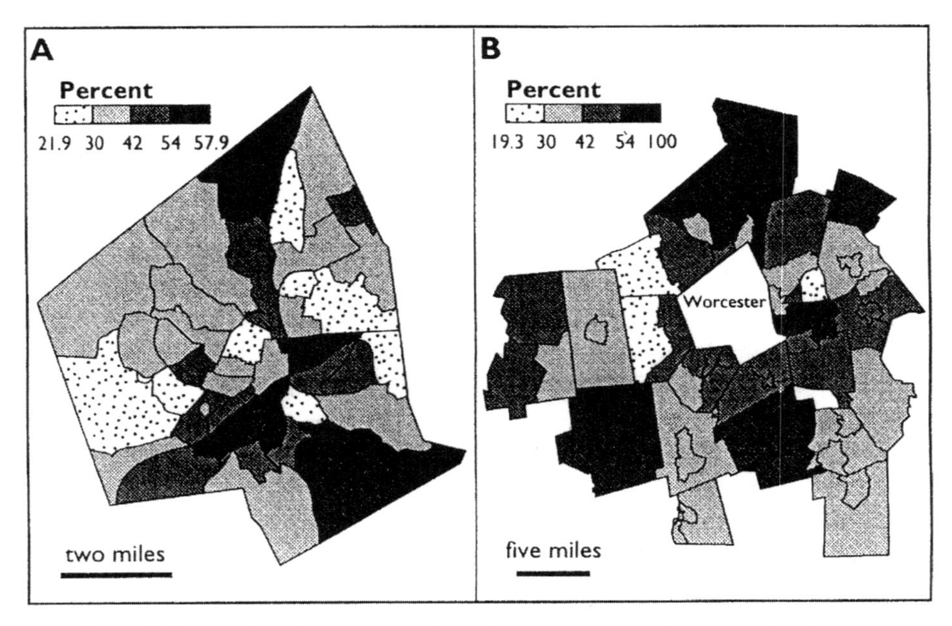

Figure 9 Proportion of employment in each census tract that is in male-dominated occupations for (A) the city of Worcester, and (B) the rest of the metropolitan area

What gendered and geographic processes help to create these maps, these very small-scale gender-specific labor markets within metro areas? I will highlight just a few of these processes here (Table 1 summarizes these processes).

1.	Home location comes first, followed by job location
2.	Women's work trips are shorter
3.	Length of work trip is related to domestic responsibilities
4.	Working part time, working in a female-dominated job, and working very close to home are all related
5.	Employers locate firms to gain access to certain types of labor
6.	Workers' and employers' search strategies help create these maps

Table 1 What processes help to create these maps?

First, almost all women and most men, too, search for a job from an already established residential location; that is, contrary to the received wisdom in urban spatial theory,[20] most people do not find a job first and then find a place to live. In our Worcester sample, which was representative of the working-age population, more than nine-tenths of the women and about two-thirds of the men reported that their residential location had been chosen before, not after, they searched for their current job. Why is this ordering of job and residential location decisions so important? Given that the geographic distribution of employment opportunities for each gender is so uneven (as you can see in Figures 8 and 9), the ordering of location decisions is important because looking for a job from an already-established and fixed residential location means that your employment options are shaped in part by where you are located on these maps that describe the geography of employment opportunity. Residential location can be particularly constraining for women because if location vis a vis potential jobs is considered in choosing a home, it is almost always the man's job, not the woman's, that enters into the location calculus.

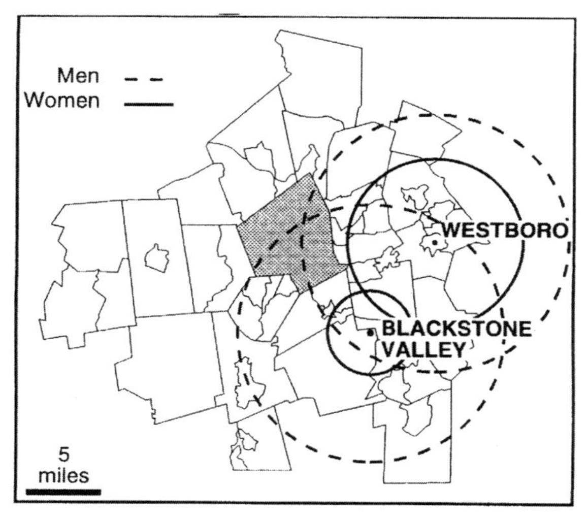

Figure 10 Median travel distances on the work trip for women and men living in two local areas of Worcester, Massachusetts

A second process that helps to explain the gendered geography of local labor markets is that on average women travel shorter times and distances to work than do men (Figure 10). This map shows the median travel distances for women and men in two suburbs of Worcester. Within the US as a whole, women's overall travel distances are shorter and their activity spaces are smaller than men's. Figure 11,

---

[20] E.g., William Alonso, *Location and land use: toward a general theory of land rent* (Cambridge MA: Harvard University Press, 1964).

which uses data from a U.S. nationwide sample, shows that women, in every age group drive fewer miles per day than men. In selecting a job, women place more importance than do men on the proximity of the job to home, and women working in female-dominated occupations are particularly likely to value proximity to home. Given a residential location that is rarely chosen with the woman's employment opportunities in mind, women's shorter commute times and distances further accentuate the importance of that residential location in shaping the employment opportunities to which a woman has access.

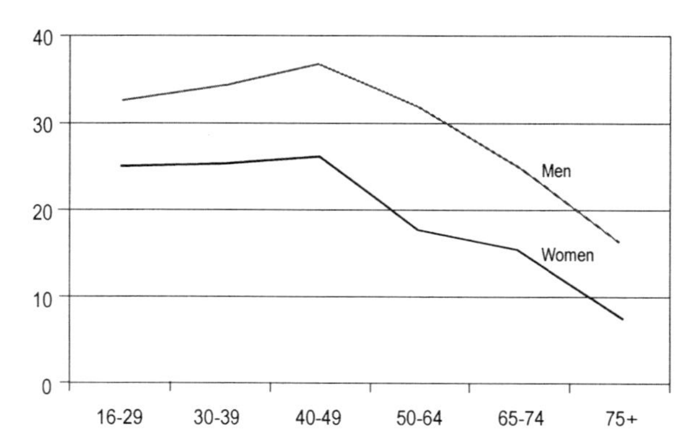

Figure 11 Women's travel distances are shorter than men's, controlling for education level (source: National Personal Travel Survey 1995)

Third, the length of the commute is related to the weight of a person's domestic responsibilities (e.g., shopping, meal preparation and cleanup, child care). People with heavier domestic responsibilities work closer to home, and this is true for both women and men. In Worcester in the late 1980s the norm was for women rather than men to shoulder most of the unpaid domestic labor in the household, which helps to explain the strong gender difference in average commute times and distances.

Fourth, being in a female-dominated occupation was not related to women's marital status or to having children, but it was related to having pre-school children and working part time. Part-time work schedules were much more widely available in female-dominated occupations than they were in other types of work. Part-time workers in female-typed jobs, moreover, had the shortest work trips and were the most likely to place a high value on having a job "close to home." Those who seek part time work and must work close to home are especially likely to work in a female-typed job, and an abundance of such jobs close to home increases the likelihood that a part-time employed mother will have a female-typed job.[21]

Fifth, employers are savvy about these facts and locate their establishments accordingly; that is, they locate so as to gain access to a particular labor force. Employers are knowledgeable urban social geographers, keenly aware of residential segregation by class, gender, and stage in the life course; they are also well aware of women's generally shorter travel times to work. As a result, they choose locations in proximity to a specific, desired labor force, which may be low-income women living in a public housing enclave, Vietnamese women living in a low-income neighborhood, or well-educated white women living in a high-income suburb.

Finally, the geography of how people search for jobs and how employers search for workers also helps to shape the patterns we saw on the maps of employment. For one thing, women and men rely on gendered social networks to find jobs, and these networks themselves have distinctive geographies; women, and especially those who find jobs in female-dominated occupations, tend to receive information about potential jobs from other women, and the jobs they learn about through these female networks tend to be close to home. In addition, employers too rely on networks to recruit workers, especially the networks of their current employees. Because such networks usually connect people who are similar (e. g., in class, race, gender), employer reliance on word-of-mouth hiring tends to reinforce existing patterns of segmentation in the labor force. In their hiring practices, employers also exhibit a preference for hiring people who live closer to the workplace over those who live farther away because they believe that this will reduce absenteeism at work, and this is particularly true in employers' treatment of women job applicants. Reflecting employers' awareness that the friction of distance tends to be stronger for women – and employers' belief that (all) women want to work close to home – some employers screen out female job applicants who they think live too far away from the workplace. The shorter work trip times and distances that we observe for women are, therefore, not simply the result of women's own choices or the spatial aspects of their personal networks; they are also the result of employers' actions as well.

These are some of the gendered geographic processes that Gerry Pratt and I learned about in our study of local labor markets in Worcester. In focusing on gender-based labor market segmentation, I thought I had finally left behind the mundane geography of everyday life and advanced to higher-level understandings of how cities are organized. But no! When we looked more deeply into what was driving these processes of labor market segmentation, what did we find? We found the geography of everyday life, that is, basically the activity patterns that had been the

---

[21] Susan Hanson, Tara Kominiak, and Scott Carlin, 'Assessing the impact of location on women's labor market outcomes: a methodological exploration', *Geographical Analysis* 29 (1997) pp. 281-97.

focus of the Uppsala study. The locations of both home and work, which are central to the geography of local labor markets, are determined in large part by the orbit that encompasses people's daily activities and by the people and places they encounter every day.

Now I want to explicitly relate each of the gendered and geographic processes I just mentioned (refer to Table 1) – the processes that help to create the maps shown in Figures 8 and 9 – to the geography of everyday life.

Let's look first at residential location. The belief is widespread that households in the U.S. are hyper mobile, that is that they change their home locations frequently. In fact, in 1990 (the most recent figures available) 80% of urban residents in the U.S. had lived in the same metropolitan area for at least five years.[22] In 1990 fully 86% of Worcester's population had lived in the area for at least five years, and the majority of the people in our representative sample (61%) had lived in the Worcester area more than 90% of their lives. This rootedness to place affects *how* people find their housing and *where* within the urban area they end up living: people who had grown up in the Worcester area (and that was the majority of people in our sample) were more likely to get their information about housing and to have found their current housing from talking with family members and co-workers than were those who had grown up elsewhere. Moreover, these long-term residents were more likely than more recently arrived people to have searched for housing in a particular small neighborhood within the metro area, a neighborhood with which they were already familiar. These findings indicate that everyday life even in the U.S. is more often sticky than it is slippery with respect to place.

Let's turn to the key findings that women's commutes are shorter than men's, that journey-to-work length is related to gender-based occupational segregation, and that the length of the work trip is related to domestic workload; these are all firmly rooted in the geography of everyday life. Here we see the power of the three principles that govern these geographies: no one has more than 24 hours in a day; no one can be in more than one place at a time; no one yet has a magic carpet that will instantaneously transport us from one place to another (we are all still subject to the friction of distance in carrying out our daily activities). Careful choice of work location – relative not only to home but also to shops, day care, and schools – can make the difference between someone being able, or not being able, to combine paid and unpaid work. Part-time work located close to home is a popular choice for mothers of young children if the woman's partner has a job with sufficient income to permit her to work part rather than full time. (Almost half (46%) of the female workforce in

---

[22] The U.S. decennial census asks if people were living in the same metropolitan area five years previously. Using data on this question for all 309 metropolitan areas in the U.S., I calculated the percentage in each metro area that were living in the same place five years ago; the average for all metro areas is 80%.

Germany work part time, compared to 26% of male workers who work part time.[23]) But working part time most often means working in a female-dominated occupation (fully two-thirds of women who worked part time in Worcester were in gender-typical jobs).

How does the geography of everyday life affect how people find jobs? More than three-quarters of the workers in our representative sample had found their jobs either by talking with people they knew or in the course of their daily activity patterns (for example, by seeing a Help Wanted poster in a window). The spatial extent of your own activity space, as well as the spatial extent of the activity spaces of the personal contacts from whom you receive job information, very much influence where you are likely to find work and how far away from home that work will be (Figure 12).

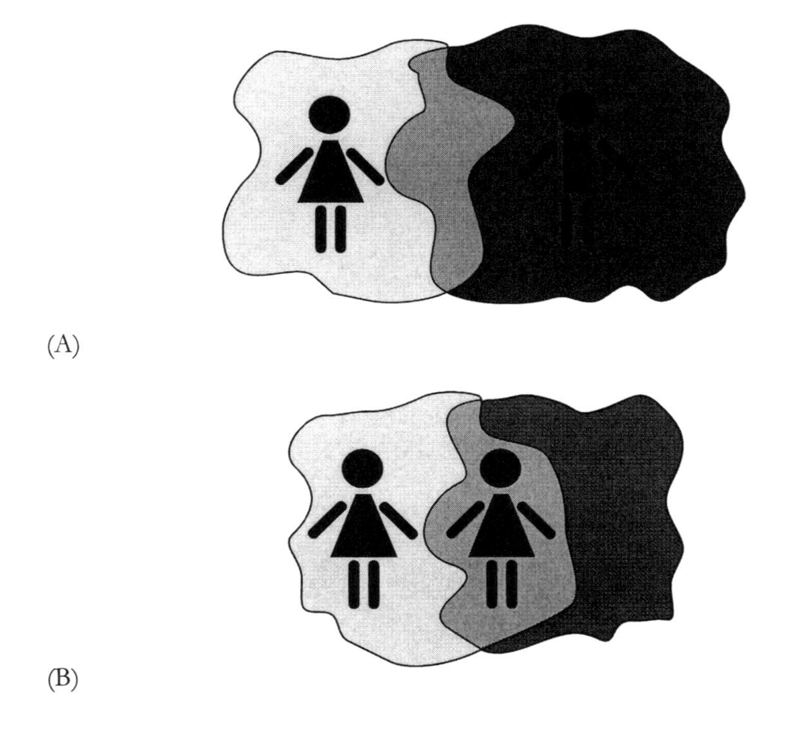

Figure 12 The size of the activity spaces of the people with whom you interact affects the geographic content of the information you receive from these people. (A) a woman receives job information from a man, whose activity space is generally larger than hers; (B) a woman receives job information from a woman, whose activity space is about the same size as her own

---

23 *http://www.destatis.de/basis/e/erwerb/erwerbtab2.htm*

Consider the woman in the diagram in Figure 12: if she receives job information from a man, whose activity space is likely to be bigger than her own, she can potentially learn about jobs over a larger area than she can from talking with a woman friend, whose activity space is likely to be about the same size as her own. People like to rely on their family, friends, neighbors, and co-workers for job information because such personal contacts are likely to know more – and have more reliable information – about a potential job than what appears in a newspaper ad.

How does the geography of everyday life affect employers' location decisions? As I mentioned, we found that employers are extremely knowledgeable about the social geography of the city; their familiarity comes largely from their own long-time residence in and travels around the metropolitan area and from talking with others. So let's not forget that employers have everyday lives, too, which very much influence their location decisions and hiring preferences.

Finally, in their reliance on current employees to recruit new workers, employers are tapping into the geography of the everyday lives of their workers, which is often the reason that a large number of workers in one workplace will come from a common residential area, even if that area is not especially close to the workplace.

These processes, which are rooted in the geography of everyday life, clearly reflect the power of the friction of distance to shape that geography. How are distance-defying communications technologies such as the cell phone and the Internet changing or likely to change the picture I've sketched out here? The real answer is that we currently know very little about how these technologies affect the geography of everyday life. But the meager evidence we have so far suggests that the friction of distance is alive and well, so that geography (including the geography of everyday life) is not dead yet (despite popular reports of the Death of Distance). One example comes from studies of telecommuting or teleworking,[24] which show that the number of U.S. workers involved in this form of work is still relatively small – on the order of 5% of the labor force[25] and that people who telecommute do so on average only about once a week.[26] These figures suggest that information technology has not yet stretched or cut that big elastic band that constrains the distance between home and workplace.

---

[24] Telecommuting refers to replacing the physical journey to work with information technology (telephone, Internet), which allows the worker to work from home or from a neighborhood telework center.

[25] Sangho Choo, Patricia Mokhtarian, and Ilan Salomon, 'Impacts of home-based telecommuting on vehicle-miles traveled: a nationwide time series analysis', Paper prepared for the California Energy Commission, 2001.

[26] Patricia Mokhtarian, 'A synthetic approach to estimating the impacts of telecommuting on travel', *Urban Studies* 35 (1998) pp. 215-41.

Another example comes from a pilot study that Sarah Niles and I did to investigate the impact of the Internet on how employers search for workers. We found that employers who advertise job openings on the Internet – which in fact means that they are casting a worldwide net in their search for workers – still had a strong preference for hiring workers who lived very close to the workplace.[27] These studies suggest that the most interesting questions concerning the impacts of information technology on the geography of everyday life are not about the death of distance but rather about the intersection of distance-defying technologies with grounded, distance-constrained everyday behaviors. It seems clear to me that the geography of everyday life may be altered by information technologies, but it will not be obliterated by them.

Before concluding this section, I should mention that the geography of everyday life has once again ambushed me in my current study on gender and entrepreneurship in Worcester and Colorado Springs. I have been surprised to find that an important reason that people start businesses has to do with their wanting to have more control over not just their lives at work but also the rest of their everyday lives – and especially the geography of their everyday lives. They want to shorten the commute (sometimes to zero by having a business at home); they want more flexibility in the times and places that they work; they want, for example, to be able to blend work and family, by having children and other family members in the workplace; they want to be able to leave work at will in order to coach a daughter's soccer team or go hiking. I had not expected to find that at the heart of entrepreneurship lies the geography of everyday life.

Conclusion

Why is the geography of everyday life so central to our work as geographers and social scientists? The research that I've been involved in indicates that people make decisions in the context of their proximate environment; others have found this to be true as well, whether the setting is an industrialized city like Worcester or a pastoral system in Burkina Faso.[28] What people see as possibilities for themselves – for example, the kind of work they aspire to and the employment opportunities they consider viable – is shaped to a large extent quite literally by what they see and with whom they interact in the course of their daily travels. The geography of everyday life is important, in addition, because people make decisions about one aspect of their

---

[27] Sarah Niles and Susan Hanson, 'The geographies of on-line job search: preliminary findings from Worcester, MA', *Environment and Planning A* (2003, in press).

[28] Valentina Mazzucato and David Niemeijer, 'Population growth and the environment in Africa: local informal institutions, the missing link', *Economic Geography* 78 (2002) pp. 171-94.

lives, such as employment, in the context of the time-space dimensions of the entire range of their own daily activities as well as those of other household members. Lifting people's decisions out of the local geographic and household context of their everyday lives leads almost certainly to flawed theories and policies.

The micro decisions that create the geography of everyday life help to explain larger-scale patterns such as labor market segmentation and the increase in entrepreneurship, especially among women. Conversely, adequate understandings of the causes and consequences of large-scale processes such as globalization require close examination of the geographies of the everyday lives of the many actors involved.[29]

The geography of local labor markets refers to the spatial variation in the nature of job opportunities (or labor market structures, which I take to include household and community structures) within a metropolitan area. I have described how the geography of local labor markets is shaped in large part through the geography of everyday life at the scale of people's daily travel activity patterns, and I have indicated how strongly gender shapes the geography of everyday life. Would we have appreciated the role of the geography of everyday life in shaping local labor markets in Worcester and patterns of entrepreneurship in Worcester and Colorado Springs if I had not been involved in the Uppsala study? I wonder. Like an older painting that is still visible even after a newer picture has been painted over it – pentimento – the Uppsala work shone through, allowing us to see that underneath the geography of local labor markets and entrepreneurship was, in fact, the geography of everyday life.

Gender not only shapes the decisions about work that I have described; gender is also created in and through these decisions, which are embedded in the totality of people's everyday lives. The contribution of feminist geography has been to demonstrate the importance of gender in the construction of distinctive spaces and places (such as local labor markets) as well as to emphasize the role of space and place in creating distinctive meanings of gender (such that, for example, a woman auto mechanic has easy access to bank loans to start a business in one neighborhood but not in another). Understanding the geography of everyday life is, then, key to understanding how different meanings of gender come to be.

Thanks to feminist analyses, it's now OK – maybe even fashionable – to speak of everyday life or the workaday world. In the 1960s – and certainly in the spatial theory-dominated discipline of geography in the 1960s – the study of everyday life was not a respectable or credible endeavor. Feminists have changed what we consider worth learning about, what we can take seriously as knowledge. My life as an

[29] Richa Nagar, Victoria Lawson, Linda McDowell, and Susan Hanson, 'Locating globalization: feminist (re)readings of the subjects and spaces of globalization', *Economic Geography* 78 (2002) pp. 257-84.

inchworm – my research adventures – have taught me that one thing we need to take seriously if we want to understand what goes on in cities is the geography of everyday life.

GEOGRAPHICAL AND FEMINIST PERSPECTIVES ON ENTREPRENEURSHIP

Geographical and feminist perspectives on entrepreneurship *

SUSAN HANSON

Introduction

I need to begin with a confession: I am not an expert in entrepreneurship and never set out to understand entrepreneurship per se. I became interested in studying self-employment and entrepreneurship through years of work on gender and local labor markets from a geographic perspective. These labor market studies prompted an interest in understanding how self-employment relates to wage and salary work. A cursory look at self-employment in the U.S. reveals a significant rise in women's business ownership since the early 1970s; this rise is so dramatic that it further increased my curiosity in, and desire to investigate, gender and entrepreneurship: woman-owned businesses now comprise more than one-third of all privately owned businesses in the U.S. (up from 5% in 1972), and the number of women's businesses is increasing at a much faster rate than is that for businesses owned by men.[1]

How does a *feminist geographer* look at entrepreneurship? I emphasize *feminist* and *geographer* because a feminist geographic perspective has been missing from the literature on entrepreneurship; furthermore, research on entrepreneurship has been absent from the feminist geographic literature on gender and work. My goal here is to provide a conceptual overview of a feminist geographic perspective on entrepreneurship and to suggest the nature of the contribution that such a perspective can make to understandings of gender, work, and entrepreneurship. A *feminist* analysis requires paying attention to gender and the power relations embedded in gender; it means taking women's lives seriously, and it also means opening up and

* I presented this material at the Heidelberg Department of Geography colloquium that was part of the Hettner Lectures, July 2002. Although I have added some references, I have tried to retain the informal tone of an afternoon colloquium. I intend this to be primarily an overview and "think piece" rather than either a detailed excavation of the literature on gender and entrepreneurship or a thorough analysis of the data on which the paper is based.

1 U.S. Bureau of the Census, *Economic Census: women owned businesses* (Washington, DC, 1972), U.S. Bureau of the Census, *Economic Census: women owned businesses* (Washington, DC, 1997); Ellen A. Fagenson and Eric C. Marcus, 'Perceptions of the sex-role stereotypic characteristics of entrepreneurs: women's evaluations', *Entrepreneurship Theory and Practice* 14 (1991) pp. 33-47; Nancy Carter, 'Entrepreneurial processes and outcomes: the influence of gender', in Paul D. Reynolds and Sammis B. White, *The entrepreneurial process: economic growth, men, women, and minorities* (Westport, CT: Quorum Books, 1997) pp. 163-77.

critically exploring categories that have long been taken for granted. A *geographic* analysis entails recognizing that no social or economic or political or cultural process is purely social, economic, political, or cultural; it is also intrinsically geographic: space and place are integral to the processes that shape entrepreneurship. A feminist geographic analysis involves examining the ways in which gender is constructed in and through the geographies of everyday life and the ways in which gender, in turn, shapes the geographies of everyday life.

In brief, a feminist geographic analysis of entrepreneurship entails a questioning and critical rethinking of most of the core concepts that comprise the study of entrepreneurship. I believe this critical rethinking is necessary if policy makers and citizens are to perceive that certain important and badly needed policy changes are possible. I seek here to provide an overview of this reconceptualization, which involves looking carefully and critically at the following questions:

- Who can be an entrepreneur? What counts as entrepreneurship?
- What is innovation? What counts as innovative?
- What is entrepreneurial context? What counts as context?
- How do we think about the motivations for entrepreneurship?
- How do we think about location decisions?
- How do we think about the relationship between entrepreneurship and place?

In my view, a re-thinking of these core concepts lies squarely at the intersection of feminist thinking and geography. The discussion will draw upon a detailed empirical study of entrepreneurship I have carried out with the help of many students in two American cities: Worcester, Massachusetts and Colorado Springs, Colorado. In this empirical work, we begin with people's lives, by listening to business owners tell their stories about how they came to start (or own) a business, how they came to locate the business where it was located, how business ownership relates to other aspects of business owners' lives including their previous employment in the wage and salary labor market, what kinds of challenges and pleasures they found in business ownership, and how their businesses relate to the places where they are located. Before taking up the set of questions that frames my feminist geographic analysis, I first briefly describe prevailing approaches to understanding entrepreneurship and then describe the design of the empirical study.

Prevailing approaches to understanding entrepreneurship

Two main approaches have dominated the ways in which scholars have thought about and sought to theorize entrepreneurship. The first, and until recently certainly the most actively pursued, focuses on *the individual* and sees entrepreneurial activity as *a quest.* The second emphasizes the importance of *entrepreneurial context.* Emblematic of the first approach is a widely used definition of entrepreneurship, which sees it as "the process by which an individual pursues an opportunity regardless of the resources he or she currently controls".[2] This view directs attention to the role of individual characteristics in determining whether or not someone will start a business. It proposes that if we want to understand the process of venture creation, we need to discover what personal traits lead to entrepreneurship. A focus on the individual and on individual characteristics (such as ambition, desire for autonomy) as the key to understanding entrepreneurship suggests that psychology is the discipline that holds the answers. Yet empirical studies fail to support the idea that personal traits alone can explain entrepreneurship.[3]

Recently, some scholars have begun to look beyond the individual and have sought to find the origins of entrepreneurship in context; they have begun to ask what contexts give rise to entrepreneurs.[4] Within entrepreneurship research, however, the words "context," "environment," and "area" have all been used in completely a-spatial and non-geographical ways.[5] Instead, scholars use "context" or "environment" to refer to the conditions prevailing in an industry (e. g., how saturated is the market for custom bicycles) rather than the characteristics of a place. One exception is Reynolds and White, who did consider the importance of community context for entrepreneurship.[6]

In the following analysis, I combine people and context, with the emphasis on gender and on place.[7] Because what follows leans heavily on what we learned from the business owners we interviewed, I next describe the study design and the nature of the empirical data bases I rely upon.

---

[2] Howard H. Stevenson and Jose Carlos Jarillo, 'A new entrepreneurial paradigm', in Amitai Etzioni and Paul R. Lawrence (eds.) *Socio-economics: toward a new synthesis* (London: Sharpe, 1991) pp 185-208, on p. 186.

[3] William B. Gartner, '"Who is an entrepreneur?" is the wrong question', *Entrepreneurship Theory and Practice,* summer (1989) pp 47-68.

[4] E.g., Claudia Bird Schoonhoven and Elaine Romanelli, *The entrepreneurship dynamic: origins of entrepreneurship and the evolution of industries* (Stanford, CA: Stanford University Press, 2001); Sarah L. Jack and Alistair R. Anderson, 'The effects of embeddedness on the entrepreneurial process', *Journal of Business Venturing* 17 (2002) pp. 467-487.

[5] E.g., Howard E. Aldrich and Ted Baker, 'Learning and legitimacy: Entrepreneurial responses to constraints on the emergence of new populations and organizations', in Schoonhoven and Romanelli, The entrepreneurship dynamic *op. cit.*, pp. 207-35.

[6] Paul D. Reynolds and Sammis B. White, *The entrepreneurial process: economic growth, men, women, and minorities* (Westport, CT: Quorum Books, 1997).

[7] By "context" I mean geographical context, or place-based context. Contextual does not equate to the local or the specific; "place" does include linkages to distant locations and place does encompass the general as well as the specific.

Research design

We interviewed owners of privately held businesses in two metropolitan areas – Worcester, Massachusetts and Colorado Springs, Colorado – that are about the same size (both about 400,000 in 1990 and both over 500,000 in 2000) but that differ in economic history, spatial structure, local culture, and population profiles. Each city is about 50 miles from a much larger metropolis, which, in each case is the state capital – Worcester is about 50 miles from Boston, and Colorado Springs is about the same distance from Denver.

An old manufacturing city located in the northeastern part of the United States, Worcester is in many ways a typical Rust Belt city, whereas Colorado Springs, located on the eastern edge of the Rocky Mountains in the American West, is emblematic of the Sun Belt. Like many Rust Belt cities, Worcester has a relatively lower-income central-city core that is surrounded by wealthier, low-density suburban communities. In Colorado Springs, the low-income neighborhoods are not concentrated in the center of the city,[8] and there is no distinct center-city/suburban dichotomy; the whole urbanized region is automobile oriented and of relatively low density. The local business culture in New England, in which Worcester is located, tends to be somewhat more formal and reliant on long-term personal relationships than does that in western cities;[9] as several business owners in Colorado Springs told us, because so many people there are recent arrivals, the hand of history does not have as close a hold on the business world in Colorado Springs as it does "back east."

This difference in the business cultures between Worcester and Colorado Springs reflects another key difference between these two places, namely the level of residential rootedness in their populations. Colorado Springs has a much higher proportion of recent in-migrants, in large part owing to the large military presence there.[10] One way of assessing the relative rootedness vs. recency-of-arrival of populations is via a question in the U.S. decennial Census of Population that asks where the person was living five years previously. In 1990, 84 percent of the people living in the Worcester metropolitan area (MSA) had been living there in 1985; the parallel figure for Colorado Springs was only 64 percent.[11] Among the business

---

8 The lowest-income households are concentrated in a few census tracts, but these tracts are not all contiguous, nor are they located in the urban core.

9 Saxenian's comparison of Boston-area and Silicon Valley computer firms describes this cultural difference; see Annalee Saxenian, *Regional advantage: culture and competition in Silicon Valley and Route 128* (Cambridge, Mass.: Harvard University Press, 1994).

10 Three large military bases and the Air Force Academy are located in Colorado Springs, employing more than 31,000 persons in 1996.

11 The comparable data available at this time from the 2000 Census refer to residence five years ago in the same *county* rather than metropolitan area. These data support the contention that Worcester-area residents are still more rooted to place than are people now living in Colorado

owners we interviewed in these two places, 90 percent of those in Worcester had lived there at least ten years, whereas only 77 percent of those in Colorado Springs had lived there that long; fully half of the Worcester business owners had lived in the Worcester area their whole lives or almost their whole lives, but this was the case for less than one-quarter (23 percent) of the entrepreneurs in Colorado Springs. For individual business people, these differences in length of residence in place translate into different levels of knowledge about tacit norms and business practices and about what kind of business is likely to be successful; at the scale of the metro area, these differences translate into different business styles, expectations, and cultures.

In each place we began with people's lives by conducting in-depth, personal interviews with a randomly selected sample of about 200 business owners; the interviews were followed by a mailed survey to a larger random sample drawn from the same target population. The data collection was designed to enable both qualitative analysis – which focuses on cases, excavates processes, and explores meaning – as well as quantitative analysis – which focuses on variables, decomposes variance, and seeks generalizations. I see these approaches as complementary rather than contradictory.

We conducted the Worcester-area interviews first, in 1998-99. Of the 200 businesses interviewed there, 45 percent were woman-owned, 42 percent were man-owned, and the rest (13%) were owned by men and women.[12] The mailed survey resulted in the return of 340 useable surveys. The Colorado Springs interviews were carried out in summer 2000; of the 180 businesses interviewed here, 41 percent were woman owned, 40 percent were man owned, and 19 percent were owned by women and men. The mailed survey in Colorado Springs yielded 504 useable responses.[13]

Springs. In 2000, 84.5 percent of the population living in the Worcester metro area had been living in Worcester County (a slightly larger area than the metro area) in 1995, whereas 66.9 percent of those living in the Colorado Springs metro area had been living in that county in 1995.

[12] Each sample was stratified by gender of business owner, as identified in the sampling frame, which was a purchased list of privately owned businesses in each metro area. The goal was to obtain a sample that was about evenly divided between male- and female-owned businesses. "Woman owned" includes businesses owned by one or more women; similarly, "man owned" includes businesses owned by one or more men. The category of "gender integrated" (or male and female owned) was created after the interviews were conducted and the actual gender of ownership had been established (in the following analysis I omit the "gender integrated" category for simplicity). After the interviews, I determined actual gender of ownership on the basis of who had launched (or purchased) the business, who was taking the risks associated with running the business, and who was making the day-to-day decisions involved in running the business.

[13] The response rates were as follows: Worcester interviews, 60.7 percent; Worcester mailed surveys, 31.8 percent; Colorado Springs interviews, 47 percent; Colorado Springs mailed surveys, 26.6 percent.

The interviews, which lasted on average about an hour and a quarter, were tape recorded, transcribed, and then coded for qualitative and quantitative analysis.[14] Semi-structured in format, the interviews posed a series of open-ended questions that asked about the start-up process ("How did you come to start your own business?"); location decisions ("How did you come to locate your business here?" with the clarification that we were interested in defining "here" in three ways – this metro area, this part of the metro area, and this building); job history (to understand how job history relates to entrepreneurship); networks of support (both personal and professional); and relationship of the enterprise to place (perceived importance to local economy, collaboration with other businesses, volunteer activities, and charitable giving).

It is important to emphasize that these businesses were selected at random from a list of privately owned businesses throughout each metro area and that, as a result, the businesses in each sample are extremely diverse in terms of size, type, and geography. The sampling strategy used in this study represents a significant departure from prior studies of gender and entrepreneurship and studies of woman-owned businesses, most of which have focused on businesses in only one or two industries,[15] often high-profile industries such as high-technology.[16] In contrast, businesses in the Worcester and Colorado Springs samples represent all industry types, including consumer and producer services, manufacturing, transportation, wholesaling, and even urban-oriented agriculture. They range in size from single-person operations (for example, a clown, stained glass artists, business consultants, someone who puts together prize packages for golf tournaments) to businesses that employ hundreds of people (e.g., temporary staffing agencies, companies providing home health care, a firm that offers employee assistance programs to other firms). In spatial extent of market area, the businesses vary from locally focused firms (e.g., hairdressers, automobile repair places, real estate companies, a medical billing firm, printers) to companies with national and international markets (a firm that markets high-technology items to scientists and engineers, an internationally active consultant on power plants, firms that focus on plastics recycling).[17] Lest we dismiss zero-

---

[14] We built a database in SPSS for quantitative analysis and used the software, NuDist 4, to build a database for qualitative analysis.

[15] E.g., Arne L. Kalleberg and Kevin T. Leicht, 'Gender and organizational performance: determinants of small business survival and success', *Academy of Management Journal* 34 (1991) pp. 136-61; Peter Rosa, Sara Carter and Daphne Hamilton, 'Gender as a determinant of small business performance: insights from a British study', *Small Business Economics* 8 (1996) pp. 463-78.

[16] E.g., Linda A. Renzulli, Howard Aldrich and James Moody, 'Family matters: gender, networks, and entrepreneurial outcomes', *Social Forces* 79 (2000) pp. 523-46.

[17] As a humorous aside, I observe that the heterogeneity of these samples signals a study designed by an urban geographer, not an economic geographer. Urban geographers tend to revel in the

employee, home-based businesses as unimportant, it is worth noting that some of the businesses with far-flung, international markets (and large gross sales) are one-person operations located at home.

Despite the great diversity of businesses included in these samples – diversity that is the direct result of the random sampling strategy employed – most of the sampled firms are small and well-established. The modal number of employees is zero (the mean is 13 employees in Worcester firms and 10 in Colorado Springs ones), the median age of companies is 9 years in each place, and the modal category for gross sales is $100,000 to $250,000 in Worcester and slightly higher ($250,000 to $500,000) in Colorado Springs.

This brief overview outlines the nature of the evidence base that I shall draw upon in developing the feminist geographic conceptualization of entrepreneurship that follows. To reiterate, the framework for this analysis involves revisiting a number of core concepts, all of which are very much interrelated:

- What is an entrepreneur? What counts as entrepreneurship?
- What is innovation? What counts as innovative?
- What is entrepreneurial context? What counts as context?
- How do we think about the motivations for entrepreneurship?
- How do we think about location decisions?
- How do we think about the relationship between entrepreneurship and place, especially the impacts of entrepreneurship on place?

In the remainder of the paper, I take up each of these questions in turn.

What is an entrepreneur?

The individual entrepreneur who has been the focus of most efforts to understand entrepreneurship, the one who is embodied in the definition mentioned earlier (an entrepreneur is someone who "pursues an opportunity regardless of the resources he or she currently controls") is, despite the mention of "he or she," very much gendered – and gendered male.[18] The discourse surrounding entrepreneurship describes the individual entrepreneur in terms that have been traditionally associated with masculinity: The entrepreneur is someone who is a risk taker, who is in control,

---

messiness of extreme diversity, whereas economic geographers usually try to constrain sample heterogeneity by focusing on one type of industry (e.g., computer manufacturing).

[18] In Gartner's review of definitions of entrepreneurship in 32 articles, I observe that whenever a gendered pronoun is used to refer to an entrepreneur, it is almost always masculine; see Gartner, 'Who is an entrepreneur?' *op. cit.*

who is independent, powerful, knowledgeable, someone who is, in brief, "a self-made man."[19]

The enduring power of this prevailing idea that entrepreneurs should be male emerged unbidden in many of our interviews. Women business owners spoke of customers who would come into their businesses and, addressing the woman who actually owned the business, ask if they might please speak with the owner (assuming that, of course, the owner could not possibly be the woman before them, but had to be a man). In some cases, women were seen as legitimate owners of certain kinds of businesses, such as beauty salons or women's clothing shops, but not of other kinds of businesses, such as trucking firms or engineering companies. Two women who had sought a bank loan with which to launch an automobile repair business told us that the banker from whom they initially sought a loan had been extremely uncomfortable with the notion that women could successfully operate a business of this type; he feared that customers would not patronize a women-owned auto shop and he denied the loan. (These women later obtained a loan from another bank and have run a very successful automobile repair business for many years.)

Against this discourse, the term "woman entrepreneur" is an oxymoron, and women entrepreneurs are considered oppositional: They are seen as contesting cultural norms and subverting traditional gender ideologies that associate femininity with incompetence, weakness, and dependence on men – ideologies that link women with only certain kinds of work or types of business ownership. In fact, many of the women business owners we interviewed *are* oppositional, in that they are deliberately crossing boundaries designed to confine women to certain domains.[20]

But the idea that all women who own businesses are doing so intentionally to subvert traditional gender ideologies ignores the structures that support and perpetuate gender differences within the labor market, including the self-employment labor market. And it ignores the fact that most people – women and men – start businesses in a field in which they have had prior experience. On the whole, women's work experiences and forms of labor market participation do differ substantially from men's;[21] likewise the domestic division of labor by and large reflects distinct gender differences, with women taking on more domestic work overall and taking care of

---

[19] This discourse seems to have saturated the thinking and research designs of most entrepreneurship researchers as well, with the result that, as recently as 1996, Starr and Yudkin observed, "We must confront the sheer paucity of data collected and research about women business owners.", Jennifer Starr and Marcia Yudkin, *Women entrepreneurs: a review of current research.* (Wellesley, MA: Wellesley College Center for Research on Women, 1996) p. 20.

[20] For a study of woman-owned businesses in male-dominated industries, see Susan Hanson and Megan Blake, 'Changing the gender of entrepreneurship', in Lise Nelson and Joni Seager (eds.) *A companion to feminist geography* (forthcoming 2004).

[21] E.g., Barbara F. Reskin and Heidi I. Hartmann (eds.) *Women's work, men's work: sex segregation on the job* (Washington, DC: National Academy Press, 1986).

different domestic tasks from men. Moreover, gendered patterns of domestic work are closely connected to gendered patterns in the paid labor market.[22]

Despite the persistence of deep and significant gender differences in the labor market, a focus on such *differences* alone glosses over the diversity of labor market experiences among women and among men; a focus on difference also ignores points of similarity in patterns of women's and men's employment.[23] In thinking about gender and entrepreneurship, we do need to recognize important gender differences in patterns of entrepreneurship and trace out the relationship of those differences to gendered patterns in wage and salary work. At the same time, we should not lose sight of the diversity of experience *within* each gender group or the similarities between women's and men's businesses; as Starr and Yudkin point out,[24] we might expect to find more differences among women business owners than between women and men entrepreneurs.

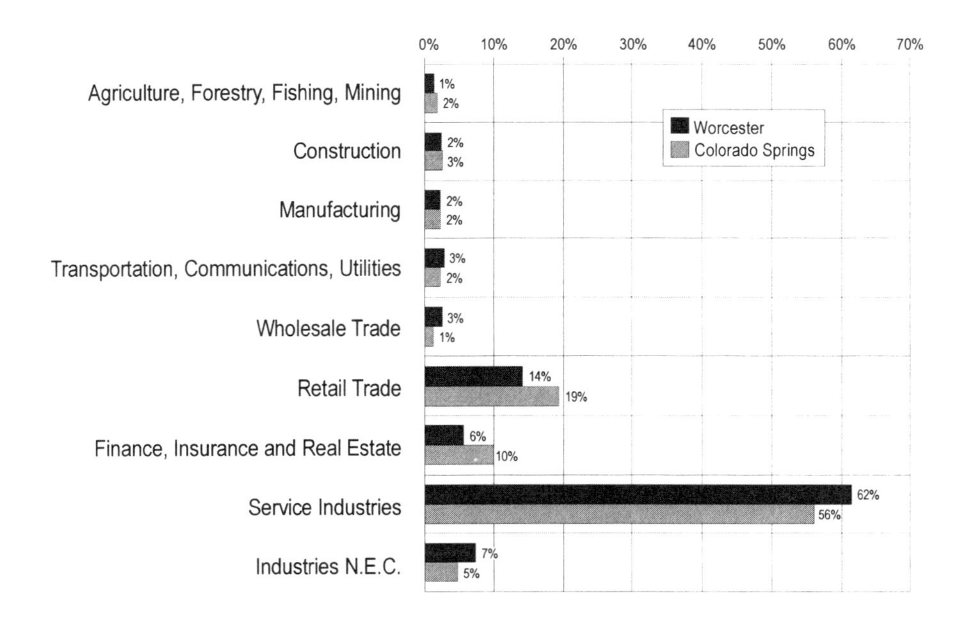

Figure 1 Percent of woman-owned businesses by industry in Worcester and Colorado Springs, 1997. Source: U.S. Bureau of the Census, 1997

[22] Susan Hanson and Geraldine Pratt, *Gender, work, and space* (London: Routledge, 1995).

[23] Kiran Mirchandani, 'Feminist insight on gendered work: new directions in research on women and entrepreneurship', *Gender, Work, and Organization* 6 (1999) pp. 224-35.

[24] Starr and Yudkin, Women entrepreneurs *op. cit.*

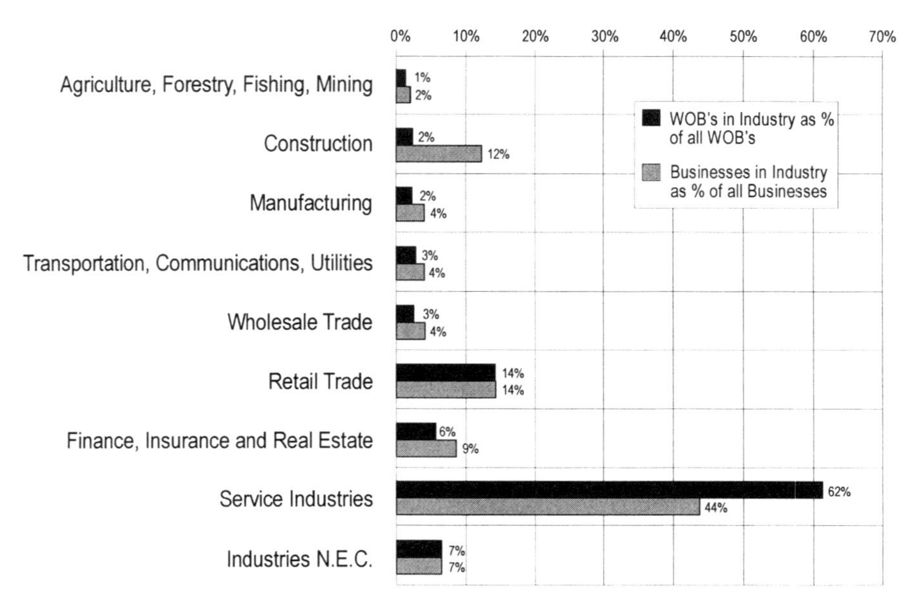

Figure 2 Woman-owned businesses vs. all firms by industry, Worcester, 1997.
Source: U.S. Bureau of the Census, 1997

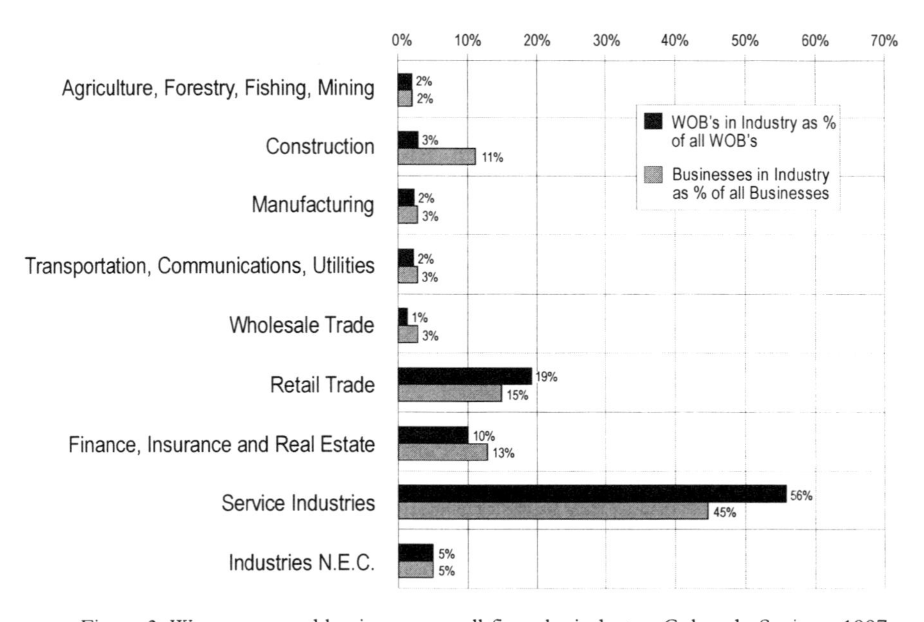

Figure 3 Woman-owned businesses vs. all firms by industry, Colorado Springs, 1997.
Source: U.S. Bureau of the Census, 1997

A number of significant differences do distinguish woman-owned businesses (WOBs) from man-owned businesses (MOBs). First, women's businesses tend to be concentrated in particular industries. Figure 1, which gives the proportion of WOBs in Worcester and Colorado Springs in each industry category, shows that WOBs in both places are predominantly in services and (to a lesser extent) retail trade. Note also the similarity of the patterns in Worcester and Colorado Springs. Figures 2 and 3 compare the distribution of WOBs among various industries to the distribution of all businesses among industries in each place. These figures reinforce the service or retail orientation of the majority of women's businesses, an orientation that is frequently noted in the literature.[25] So, for example, whereas 44 percent of all businesses in Worcester are in the service industry, fully 62 percent of all WOBs are in services (Figure 2). The representation of WOBs in the retail trade in Worcester is in proportion to the representation of retail among all businesses (14 percent), but WOBs are underrepresented in all industries other than services and retail trade. They are especially poorly represented in construction, which comprises 12 percent of all businesses but only 2 percent of all WOBs. Comparison of Figure 2 and Figure 3 reveals that a higher proportion of WOBs in Colorado Springs than in Worcester are in retail trade, a finding that probably reflects the greater importance of tourism to the Colorado Springs economy.[26]

A second dimension that distinguishes women's from men's businesses is that WOBs are generally smaller than MOBs as measured by number of employees and by gross sales.[27] In both Worcester and Colorado Springs MOBs were more likely than WOBs to have any regular employees at all, and the gender difference is more pronounced in Worcester, where 87 percent of MOBs but only 60 percent of WOBs had any regular employees. In Colorado Springs the comparable figures were 76 percent of MOBs and 70 percent of WOBs. The median number of employees was four for MOBs in both places, but for WOBs it was only one in Worcester and two in Colorado Springs. The gross sales of women's businesses are significantly lower than are those of men's businesses. In Worcester, nearly half (47 percent) of businesses owned by women had annual gross sales of less than $100,000, but only 5.4 percent of male-owned businesses had gross sales below this level; at the other extreme, half of men's businesses, but only 15.7 percent of women's businesses had gross sales that exceeded $500,000. This general pattern, which is widely documented in the literature, also holds for Colorado Springs.

---

[25] Ibid.

[26] Many of the woman-owned retail businesses in Colorado Springs are oriented to tourists.

[27] Carter, 'Entrepreneurial processes and outcomes' *op. cit.*; Starr and Yudkin, Women entrepreneurs *op. cit.*

Third, and perhaps most interesting for geographers, gender differences are evident in some geographic dimensions of entrepreneurship. In general, women tend to be more local in their business orientation. For example, women were less likely than men to have considered starting their businesses outside the Worcester area; only 11.5 percent of the women we interviewed, but 29 percent of the men, said they had considered starting their businesses elsewhere.[28] This greater propensity of women to favor local startup may be due to another finding, namely that women business owners in both places had lived in the local area significantly longer before starting a business than had their male counterparts;[29] this greater residential rootedness among women may lead to higher levels of familiarity with place and may contribute to women's greater disinterest (than men's) in moving elsewhere to start their businesses. Women's businesses were also located closer to their homes than were men's.[30] Women were not only less likely than men to consider locating a business far from home, considering either the Worcester metro area relative to the rest of the world or locations within the Worcester area; they were (in Worcester, but not Colorado Springs) far more likely than men to locate their businesses *in* the home.[31] Another way in which women's businesses were more locally oriented than men's businesses is in the spatial extent of their market areas. On average, 88 percent of the domestic sales (that is, sales within the U.S.) of woman-owned businesses vs. 77 percent of the domestic sales of male-owned businesses were within the Worcester metro area.[32]

Lest a focus on gender differences obscure other important points in the previous paragraph, it seems worth pausing to recognize that, overall, a relatively small proportion of business owners of either sex consider locating their businesses

---

[28] ($p<.01$). The comparable figures from the Colorado Springs mailed survey are 16.5 percent of WBOs and 25 percent of the MBOs ($p=.03$).

[29] In Worcester, women business owners had lived in the local area (defined as the MSA) an average of 26.1 years before starting a business, compared to 20.1 years for male business owners ($p=.02$). In Colorado Springs, the comparable figures are 13.6 for women vs. 8.3 for men ($p=.02$).

[30] This gender difference is significant in Worcester but not Colorado Springs: average travel time between home and business was 10.7 minutes for women vs 14.3 minutes for men in Worcester ($p=.03$); in Colorado Springs it was 13 minutes for women vs 16.4 for men (ns).

[31] Among the woman-owned businesses interviewed in Worcester, 30.7 percent were located at home, compared to 8.5 percent of MOBs. We found no gender difference, however, in the proportion of businesses that had once been located at home; overall, 18 percent of the Worcester businesses interviewed were located at home and an additional 18 percent had once been located at home but were located outside of the home at the time of the interview. The gender difference in the proportion currently at home, together with the lack of gender difference in the proportion ever located at home, suggest that men and women are equally likely to start a business at home but MOBs are more likely than WOBs to move their businesses out of the home.

[32] ($p=.02$). We found a similar gender difference in Colorado Springs in percentage of domestic sales made within the local metro area (78 percent for WOBs vs 69 percent or MOBs), but the difference there is not large enough to be statistically significant.

anyplace other than the place where they are currently living. Within that place, people locate their businesses relatively close to their residences; in fact, most people say that their commutes are shorter after starting a business than they were before. Finally, although the market areas of WOBs are more localized than are those of MOBs, the average percentage of sales within the local metro area ranges from three-fifths to three-quarters for all gender ownership types, indicating that sales are biased toward the local regardless of gender of owner. Sometimes larger points like these get buried in a discussion focused on gender differences.

Also often masked in gender comparisons are the similarities between women's and men's businesses. For example, although in both Worcester and Colorado Springs woman-owned firms were more likely than their man-owned counterparts to be an owner's first business, we found no gender difference in age of the businesses studied. The median age in both places was nine years. More interesting is the finding that women and men were equally convinced that their businesses were successful;[33] 94 percent of those interviewed in Worcester[34] and 97 percent of those interviewed in Colorado Springs said they thought their businesses were successful.[35] Perhaps this widespread perception of success explains the absence of another gender difference: roughly equal proportions of women and men in both places – and relatively few business owners in each place – said they would consider moving their businesses outside the local metro area.[36]

In concluding this section, I want to emphasize three points in asking, "Who is an entrepreneur?" – points that future studies should address. First, I want to question the association of entrepreneurship with men; as I mention in the next section, this association has proven pernicious for women business owners and aspiring women entrepreneurs. Second, I want to open up the category "entrepreneur" to a gendered analysis; this involves exploring gender differences and similarities as well as the great diversity within WOBs and MOBs, diversity that I have glossed over in this discussion. Such a gendered analysis needs to ask whether there is more variation among WOBs than between WOBs and MOBs and needs to understand these patterns in the context of structures both in and out of the labor market that shape the processes of gendered entrepreneurship. As Mirchandani has pointed out, we need to "focus on gender as a process [that] is integral to business ownership, rather

---

[33] We asked, "Do you think your business is successful?" and followed that question with "What is your meaning of success?"

[34] Only 1.5 percent said they thought their businesses were not successful.

[35] In a later section of this paper I describe some interesting gender differences in the meaning of "success."

[36] 14 percent in Worcester and 19 percent in Colorado Springs said they would consider moving their businesses outside the local metro area.

than as a characteristic of individuals."[37] Third, I want to point out that who can legitimately be considered an entrepreneur varies from place to place; the prevailing ideology and the gatekeepers (such as bankers, lawyers, and business advisers) in some places consider female ownership of any business or of certain kinds of businesses to be inappropriate and hence unworthy of support, but in other places such gatekeepers will help facilitate rather than constrain aspiring women entrepreneurs.[38]

What is an innovation?

Closely related to who counts as an entrepreneur is the question of what counts as innovation. A long and intense debate has focused on the distinction between true entrepreneurship (which allegedly involves innovation, business growth, and wealth creation – i.e., increasing profits) and mere self-employment (which some see as simply another form of labor force participation, an alternative to wage and salary work, but not innovative, not wealth creating).[39] Gartner and Light and Rosenstein, among others, have criticized this distinction, pointing to the difficulty of measuring innovation and calling for a broadly inclusive definition of entrepreneurship.[40] I follow Lavoie,[41] in considering as an entrepreneur anyone who has taken the initiative of launching a new venture, who is accepting the associated risks and the financial, administrative, and social responsibilities, and who is effectively in charge of its day-to-day management. This definition avoids making the invidious distinction between true entrepreneurship and mere self-employment.

The distinction, based as it is in "innovation," reminds me of the line drawn between basic and non-basic economic activities that is a staple of introductory economic geography. Recall that basic economic activities lead to exports; they are valued because they bring income to a place from other places. By contrast, non-basic economic activities serve only a local market and are therefore viewed as not

---

[37] Mirchandani, 'Feminist insight on gendered work' *op. cit.*, p. 230.

[38] One example is the women who were able to get funding and start a thriving auto repair business in one part of the urban area, after being denied a bank loan in a different part of the Worcester region.

[39] See, e.g., Robert L. Aronson, *Self-employment: a labor market perspective* (Ithaca, N.Y.: ILR Press, 1991).

[40] Gartner, 'Who is an entrepreneur?' *op. cit.*; Ivan Light and Carolyn Rosenstein, 'Expanding the interaction theory of entrepreneurship', in Alejandro Portes (ed.) *The economic sociology of immigration: essays on networks, ethnicity, and entrepreneurship* (New York: Russell Sage Foundation, 1995) pp. 166-212.

[41] Cited in Dorothy P. Moore, 'An examination of present research on the female entrepreneur – suggested research strategies for the 1990's', *Journal of Business Ethics* 9 (1990) pp. 275-81, on p. 276.

income generating and consequently as relatively unimportant to the economic competitiveness of a place. Innovation is prized by economic geographers and local economic developers alike because it is linked to exports, to impacts beyond the local place, and therefore to profits for the place of origin.

I believe that we should think about innovation and innovative economic activity in terms of local impacts not just extra-local impacts. Valuing the positive contributions that products and services make to the life of a community brings into focus the local impacts of many businesses that are innovative in terms of meeting local needs and creating viable communities. These kinds of innovative businesses are usually dismissed as irrelevant to economic growth because they do not serve extra-local markets or attract large profits from distant places. Yet such locally focused innovations do increase the attractiveness of a community, including its attractiveness to outside investment, which some may view as the only positive outcome of innovations aimed at meeting local needs.

Dozens of the business owners we interviewed have created ventures that involve important local innovations that would most certainly not count as innovative in the traditional literature. They would not be considered innovative because their innovations are not technological and are not oriented toward generating exports. Their innovations are, moreover, distinctly geographical, in that their innovativeness depends on geographical context; they create something that is new *to that place.* Consider a few examples.

A woman we interviewed in Worcester parlayed experience as a nurse into a successful business venture by converting her large home into a respite care facility.[42] She had been working in a nursing home while caring for her aged mother at home and was keenly aware of the lack of any facility like the one she started in her town: "We had my mother living with us for many years, and we didn't have any – there was no place to turn to so that we could go out. So we never went anywhere or did anything. I'm a nurse, and the more I was out there seeing people putting [their relatives] into nursing homes because they couldn't have a life, the more I put my life together with theirs, and I thought there must be a need for it. So this [her respite care facility] allows people to keep their loved one at home and yet have a life of their own. " The addition of this woman's business contributes substantially to the quality of life of many families living in the Worcester metro area.

Many businesses, like the respite care facility, fill a community need within a specific geographic area. The people who launched these firms recognized a spatially defined need and filled it. Many made remarks like this one, from a woman who runs a taxi service in a low-density, low-income suburb: "Without this business, people in

[42] A respite care facility offers brief respite (from part of a day to several weeks) for people caring for others (e.g., a handicapped child or an elderly relative) in their homes; without such facilities, caregivers often never have even an evening to themselves, much less a vacation.

this town who don't have cars are stuck; they have no way to get to the doctor, no way to get to the store." Similarly, a woman who started a children's clothing consignment shop in a low-income suburb said, "People here don't have much money to spend on their kids' clothes. If I weren't here, most of them would not have access to cheap clothes for their kids." Another entrepreneur who identified a spatially defined need and filled it is a Hispanic man who started a supermarket in his low-income inner-city neighborhood, where most people walk or take a taxi to the food store because they do not own cars and where, for decades, there had been no store selling fresh fruits, vegetables, or meats.

A final example of an unusual, but in my view important, type of innovation comes from a woman who owned a diner.[43] Early in the interview this woman described her view of the role of diners in the life of a community: a diner is a place where single people can go to have a simple, healthy, inexpensive meal and enjoy the company of others (a diner has a long counter, along which customers sit on stools). She pointed out that diners have traditionally been male owned and that traditional diner culture has made diners places where male customers feel they can harass women, whether those women are other customers, waitresses, or cooks. She said that after she took ownership of this diner, she had made it clear to her male customers that she would not tolerate any mistreatment or harassment of women; she insisted that her women customers should feel safe and comfortable in her establishment. This woman with a tenth-grade education and a long history of working as a waitress or a food preparer in other restaurants, said, in response to our question about what the owner sees as innovative about her business,[44] that she had created a woman friendly diner. This may not be the kind of wealth-creating innovation that economic geographers are usually interested in, but I think it illustrates the kind of place-transforming innovation that many entrepreneurs are creating.

Because a higher proportion of women's than of men's businesses are focused on local markets, acknowledging the importance of local impacts opens the possibility of recognizing the innovative capacity of many women's businesses that are currently viewed as "merely" an alternative form of employment for their owners. Such a shift in worldview could change how community gatekeepers treat aspiring and nascent women entrepreneurs. Most local economic development policy currently seeks to attract and support mainly basic economic activity, and local gatekeepers' actions support this policy. If bankers, for example, were to perceive women's proposed/

---

[43] Diners are a distinctly American form of eating establishment. Originating near Worcester, the diner began as a railroad dining car parked along the side of the road. "Classic" or vintage diners now have great appeal as tourist attractions in the U.S. Northeast.

[44] We had to explain the meaning of the word "innovative" to her; we indicated that we wanted to know if there was anything that made her diner unusual or different from others.

existing businesses as potentially innovative rather than dismissing them as simply non-basic, they would be less likely to deny women access to the resources needed to launch or expand a business. Such a shift would be one important policy outcome resulting from a change in thinking about who can be an entrepreneur and what counts as innovation.

In closing this section, I want to highlight three main points about how we think about innovation. First, we need to think geographically about how innovations are defined; what defines an activity as innovative is the geographic area in which that activity is introduced: something is innovative within a particular geographical context. Second, we need to expand the idea of what counts as innovative economic activity beyond the technological; many non-technological activities create value. Finally, we need to consider the local as well as the non-local impacts of economic activity and appreciate as truly innovative those locally focused activities that introduce something that is new to a place and that significantly changes life in that place.

Context

Scholars of entrepreneurship are increasingly recognizing that the conventional understanding of entrepreneurs as *individual* achievers, loners, and people who "make it on their own" does not accurately capture the actual process of venture creation.[45] The importance of context to the startup process and to the process of running a business has increasingly become the focus of attention. As Reynolds and White wryly note, contexts do not start businesses, people do;[46] yet to ignore the contexts within which people act is to disregard a variety of significant environmental influences that enable and constrain entrepreneurial activities, influences that furthermore are susceptible to policy intervention.

As I mentioned earlier, when a geographer hears "context," he or she automatically thinks "geographic context," but for entrepreneurship researchers, "context" or "environment" do not first and foremost refer to place; instead, these words refer to the conditions within an industry, such as the price or demand and supply structures within an industry.[47] The unspoken scale is national or global, not

[45] Reynolds and White, The entrepreneurial process *op. cit.*; Schoonhoven and Romanelli, The entrepreneurship dynamic *op. cit.*

[46] Reynolds and White, The entrepreneurial process *op. cit.*

[47] E.g. Kalleberg and Leicht, 'Gender and organizational performance' *op. cit.*; A. Lomi, 'The population ecology of organizational founding: location dependence and unobserved heterogeneity', *Administrative Science Quarterly* 40 (1995) pp. 111-44; M. Low and E. Abrahamson, 'Movements, bandwagons, and clones: industry evolution and the entrepreneurial process', *Journal of Business Venturing* 12 (1997) pp. 435-57. I am aware of two exceptions to this generalization. In

local. Some entrepreneurship researchers use "context" to refer to social context, but again they are referring to national[48] or organizational[49] structures, not local ones. The global or national scales are not, however, the scales at which most businesses are launched. Most businesses start small, in the place where the owner-to-be lives. It is therefore crucial to understand how geographic context at regional, metropolitan, and sub-metro scales affects entrepreneurship. We have already seen that geographic context influences who is viewed as a legitimate entrepreneur for what kind of entrepreneurial activity and what is considered innovative.

Geographers conceptualize context at varied spatial scales; among the dimensions of context at regional, urban, and sub-metropolitan scales that seem relevant to understanding entrepreneurship are the following:

- Industry and occupation structures; for example, is local industrial structure an oligopoly or made up of many small businesses?[50]
- Formal institutions like laws and organizations (examples include zoning laws; Chambers of Commerce and other business organizations like the Better Business Bureau; banks; other firms);
- Informal institutions (e.g., social norms and values, labor skills, patterns of volunteering, social networks);
- Spatial structures (the relative location of a metro area; land use and transportation patterns, traffic congestion levels); and
- Demographic patterns (age and education structures; rates of in-migration and out-migration).

In addition to geographic context, family and household context are also important to the decision to start a business. Among the factors to consider here are the following:

---

one, Reynolds and White (The entrepreneurial process *op. cit.*) asked a random sample of respondents from Wisconsin to rate aspects of their local communities in terms of the importance of each to their businesses; Reynolds and White found that their sample of business owners considered access to customers, availability of capital, and quality of life more important than business costs. The second is an in-depth qualitative study of seven business owners in Scotland, which focused on the individual's relationships within the community; see Jack and Anderson, 'The effects of embeddedness' *op. cit.*

[48] Johann Peter Murmann and Michael L. Tushman, 'From the technology cycle to the entrepreneurship dynamic: the social context of entrepreneurial innovation', in Schoonhoven and Romanelli, The entrepreneurship dynamic *op. cit.*, pp. 178-206.

[49] Aldrich and Baker, 'Learning and legitimacy' *op. cit.*

[50] B. Chinitz, *Economic study of the Pittsburgh region* (Pittsburgh, PA: Pittsburgh Regional Plan Association, 1961).

- Having parents or other significant relatives who were entrepreneurs increases the probability that someone will eventually become self-employed;
- Family often contributes capital to the nascent entrepreneur, not just financial capital but also ideas, labor, and – importantly – encouragement; the location of family members (nearby or distant) can influence the kind of capital family members contribute;
- Having an employed spouse or partner is especially important, not only as a source of startup capital, but also as a source of a steady income and employer benefits; an employed spouse or partner is a kind of safety net for someone launching a business;
- Many nascent entrepreneurs use the family home as collateral against which to borrow start-up funds;
- Having children at home is positively related to being self-employed for women;
- Family can also be a deterrent to business startup, as in the case of people who wait until their children are grown before starting a business.

It is the individual's relationship to these contexts – geographic and familial – that is important in shaping entrepreneurship. Particularly important in my view is the embeddedness of the individual in the local and regional context.[51] Such embeddedness, or rootedness to place, yields place-specific knowledge (knowledge of market potential, good possible business locations, labor characteristics, local norms, knowledge of what business ideas and practices will and will not fly in this place); it also yields local reputation and trust. In short, being embedded results in "knowing and being known" as Hart *et al.* have put it,[52] which, for the potential business owner, significantly reduces search costs and reduces risk. A key point here for a geographer is that these benefits that accrue from being embedded in a place are geographically fixed assets; they are not movable or transferable to some other place.[53]

To conclude this section on context, I recount the story that two business co-owners in Worcester told us about how they had come to start their firm, which is a

[51] See also Jack and Anderson, 'The effects of embeddedness' *op. cit.*

[52] Myra M. Hart, Howard H. Stevenson and Jay Dial, 'Entrepreneurship: a definition revisited', in William D. Bygrave *et al.* (eds.) *Frontiers of entrepreneurship research 1995* (Wellesley, MA: Babson College, 1995) pp 75-89, on p. 83.

[53] It is interesting in this regard to note that only a minority of respondents in Worcester and Colorado Springs said that no one had been helpful to them in the startup process (this was in response to a question that asked, "Were there any people who were really important to you in your decision to start your own business or were helpful to you in the startup process?"). Striking gender differences are evident in the proportions that claimed to have had no help: whereas 24 percent of the men in both places said that no one had helped them (they had done it by themselves), only 4 percent of the women in Worcester and 8 percent of the women in Colorado Springs could identify no one who had been helpful in the start-up process.

business that inspects shoes produced outside the U.S., mainly in Asia. Each of the co-owners, a man and a woman now both about 60 years old, had worked for decades in the Worcester shoe industry, once one of Worcester's thriving industries. Each is a lifelong Worcester resident with a high-school education and additional education acquired via night school. Each had spent many years working in local shoe factories, and before starting this business, the man had been head of quality control at one of the local shoe companies. Most shoe companies in Worcester closed down in the late 1980s and early 1990s, leaving a large labor force of primarily female skilled shoe workers without jobs. Some of these unemployed women approached the two current owners and asked them to start something that would employ shoe workers locally. In the interview the current owners spoke repeatedly of the loyalty among "shoe people" and their own sense of responsibility to the people with whom they had worked for decades. They started this business, which involves inspecting shoe products made elsewhere, because they knew, from their years of work in the shoe industry, that quality assurance was a needed niche. In his position as head of quality control for one of the firms that had closed, the man had traveled widely and gained knowledge of the need for quality control on imported shoes. Their firm now employs about 15 women full time and an additional 45 on an as-needed basis. Despite the importance of local ties to the inception of this business, all of their market is international; none is local.

This example highlights the importance of context in several ways, including the individual's rootedness in, and commitment to, place and the industrial history of a place, which in this case led to a culture of "shoe people helping shoe people." The social networks these entrepreneurs drew upon in launching their business spanned local and international scales. In terms of family context, both owners were married (not to each other) and had the support (financial and otherwise) of their spouses in launching the business; moreover, all of their children were now grown, so that they felt they could take the risk of starting a business.

Motivations for starting a business

The literature on motives for entrepreneurship highlights a desire for autonomy as the most important reason people choose self-employment over wage and salary work,[54] and studies show no significant gender difference here: women and men alike cite a desire for greater independence as the main reason for starting a business.[55] But

[54] Aronson, Self-employment *op. cit.*; Reynolds and White, The entrepreneurial process *op. cit.*

[55] Starr and Yudkin, Women entrepreneurs *op. cit.*; Sara Carter and Tom Cannon, *Women as entrepreneurs: a study of female business owners, their motivations, experiences, and strategies for success* (New York: Academic Press, 1992). It is interesting to note that, according to Reynolds and White (The

I wonder what "a desire for autonomy" really means and how this concept has been measured in entrepreneurship research; have people simply checked "autonomy" on a list of possible reasons for launching or owning a business? As Marlow and Strange point out, research on entrepreneurship has been largely gender blind, and the lack of a gendered analysis assumes that women entrepreneurs have similar motivations for, and are looking for similar rewards from, business ownership.[56] The data from Worcester and Colorado Springs reveal interesting patterns, some distinctly gendered, in motivations for startup and expectations from self-employment.

The aspiration for greater autonomy in the workplace did indeed emerge as an important motivation for business ownership among the people whom we interviewed. In response to the open-ended question, "Could you trace out how you came to start your own business?" 35 percent of entrepreneurs in Worcester and 41 percent of those in Colorado Springs voiced a desire for greater independence as a motive for launching a business.[57] In the discussions that followed this open-ended question, business owners described a wide variety of motivations for launching a business (by the time we had finished coding the interviews, we had created 61 different codes to capture these, ranging from "friend encouraged me" to "love the work"; "autonomy" was one of these 61 codes). One motivation that seems related to autonomy was the desire to achieve a more satisfactory "work-family balance"; 13 percent of the women business owners we interviewed in Worcester spontaneously mentioned this as a reason for launching a business.

Because this motivation emerged so frequently in the interviews, we included a specific question about it in the mailed survey. We asked respondents to indicate how much they agreed with a number of statements, one of which was, "I wanted to have my own business so I could have more flexibility for family responsibilities"; about two-thirds of women and men either "agreed completely" or "generally agreed" with this statement (64 percent of men and 67 percent of women in Worcester; 63 percent of men and 61 percent of women in Colorado Springs). Women, however, were much more likely than men to say they "agree completely" with this statement: 45 percent of the women vs. 24 percent of the men did in Worcester; this contrast is

---

entrepreneurial process *op. cit.*), motives for business startup do not appear to be related to a person's attitudes toward risk.

[56] Susan Marlow and Adam Strange, 'Female entrepreneurs: success by whose standards?', in Morgan Tanton (ed.) *Women in management: a developing presence* (New York: Routledge, 1994) pp. 172-84.

[57] Significantly, even though autonomy is clearly important, it was mentioned by a minority of the business owners with whom we spoke. In both places a higher proportion of men than of women mentioned autonomy, but in Colorado Springs the gender contrast was notable, with half of the men, compared to one-third (34 percent) of the women including a reference to autonomy in their response. In Worcester, the comparable figures were 38 percent of men and 32 percent of women.

more muted[58] in Colorado Springs: 36 percent of women vs. 27 percent of men. What do these figures really mean? What do people talk about when they say they are searching for a better balance between family and work, between caring work and paid work? It turns out that they are talking about the geography of everyday life.

How so? How does entrepreneurship relate to the geography of everyday life? What strategies for promoting a better balance between work and family does entrepreneurship allow? Among the strategies people mentioned were the following: (1) One widely adopted approach is reducing the length of the commute and thereby creating a more localized life. In the mail surveys 56% of business owners in each place said their commute was now shorter than it had been before starting a business; as noted above, sometimes this means reducing the commute to zero by basing the business at home. (2) For some owners, the autonomy associated with running a business translates into the ability to reduce or eliminate non-local business travel; indeed some cited excessive out-of-town travel in previous jobs as an important motivation for entrepreneurship. (3) Business ownership allows more flexible hours of work (not necessarily fewer hours of work) and more control over when and where one works.[59] [60] A related strategy involves consciously constraining the size of the market area and limiting the size of the business or expanding and contracting the business as family demands wax and wane.[61] (4) Business ownership also allows people to blur the boundaries between family and work by hiring family members, caring for children and elders in the workplace, or having a business that closely relates to a spouse or partner's work. Finally, (5) by creating a job for themselves "here," starting a business enables some people to remain in the place where their family's roots are rather than having to move elsewhere for work.[62] In sum, many of the motivations people mentioned for launching a business have to do with gaining

---

[58] (but still statistically significant at $p=.04$)

[59] Another question in the mailed survey asked respondents how much they agreed or disagreed with the statement, "I have more control over my time now"; in Colorado Springs nearly three-quarters (73 percent) either "completely agreed" (42 percent) or "agreed" (31 percent) with this statement (no gender difference). The comparable figures for Worcester were almost identical: 72 percent either "completely agreed" (41 percent) or "agreed" (32 percent) with the statement.

[60] A desire for greater flexibility in the time-space management of daily life was mentioned by many of our respondents who had a background in the corporate world; from the Worcester interviews, 8 of the 13 women and 5 of the 16 men who had worked for large companies mentioned this as a reason for leaving the corporate world to start a business of their own. For these people, it was not the lack of opportunity (or "glass ceiling") that made them refugees from corporate America; it was the desire for more flexibility in their lives.

[61] From the mailed surveys we learn that 40 percent of women and one-third of men had cut back on business for family-related reasons in Worcester; the comparable figures for Colorado Springs were 30 percent of women and 31 percent of the men.

[62] This strategy is more prevalent in Worcester (than in Colorado Springs) because of the greater residential rootedness of the population there.

more control over daily travel-activity patterns, over the time-space dimensions of daily life; these motivations are intimately related to the geography of everyday life.[63] Moreover, they are salient, while not necessarily equally so, for men and women alike.

These reasons for startup are closely related to another, unexpected, motivation for starting a business, one that is probably also related to "autonomy," namely the desire to have a workplace where one will actually enjoy working, or as several people put it, "to create a workplace where I want to work." Because this motivation emerged spontaneously from the Worcester interviews, in the Colorado Springs interviews we specifically asked, "Was creating a better workplace one of your motivations for starting your own business [becoming self-employed]?" To this question, 60 percent of men and 69 percent of women responded in the affirmative. A significant part of creating a workplace where one will enjoy working is evident from the interviews: these entrepreneurs had a passion for what they were doing, whether it was golfing, sewing, roofing, interior design, horses, dogs, plants, or automobiles (to name but a few of the passions the entrepreneurs in our study were pursuing); for them, creating a business was a way to devote their working time to something they truly enjoyed doing. This kind of intense interest does not preclude an interest in making a profit,[64] but it does create a working life that is more meaningful than working in another field.

I have touched on only some of the motivations driving venture creation and have emphasized autonomy because it is the factor most widely cited in the literature as an explanation for self-employment. Looking inside the often-reported category "autonomy" as a motivation for entrepreneurship reveals considerable complexity. Much of this complexity is related to the geography of everyday life (e.g., having control over when and where one works), and much of it is also related to creating meaning in people's lives (e.g., creating a workplace where one can enjoy working; pursuing a deep interest in a particular line of work). In addition, a look at what autonomy actually means reveals the importance of family and community in people's reasons for entrepreneurship; people want to create lives in which paid work

---

[63] In the Colorado Springs interviews we specifically asked if people had been able to achieve a better balance between work and family by owning a business; 47 percent of the women and 40 percent of the men responded affirmatively (about 9 percent of each group said no; the rest indicated "somewhat" or "yes, but not a goal").

[64] In fact, for the majority of our respondents, the meaning of "success" is financial, although there are interesting differences between the two places and between women and men in the meaning of success. In the interviews, we asked, "Do you think your business is successful?" (as reported earlier, almost everyone said "yes"), and that question was followed immediately by "What is your meaning of success?" A higher proportion of business owners in Colorado Springs (68 percent) than in Worcester (59 percent) mentioned profits or money in response to this open-ended question, and in both places men were more likely than women to say that success means financial success: 75 percent of men vs. 44 percent of women in Worcester and 77 percent of men vs. 61 percent of women in Colorado Springs.

– even meaningful paid work – allows a place for family and community involvement.

Location, location, location

Very few entrepreneurship scholars consider location, and none that I am aware of has examined the gender dimensions of entrepreneurial location. Economic geographers, who do consider location, like to think that entrepreneurs select locations for their firms after scanning the nation or the globe and identifying places that have ideal locational attributes for their businesses. In their nationwide survey of entrepreneurs, however, Reynolds and White[65] found that most people simply launch their businesses in the place where the owner had lived; that is, most people do not engage in an extensive search process and then move somewhere new to start a business. The Worcester and Colorado Springs data support the idea of indigenous startups and reveal significant gender dimensions to the location process.

In the interviews, we asked, "How did you come to locate your business here? By "here" we mean the Worcester [Colorado Springs] metro area, this part of the Worcester [Colorado Springs] area, and this building." The most common response to this open-ended question was "because this [metro area] is where I live" or, to a lesser degree in case of Colorado Springs, "I knew I wanted to move (back) here."[66] In fact, many respondents were quite puzzled as to why we were even asking such an "obvious" question as why they located their businesses "here"! The majority of the people we interviewed, even in Colorado Springs where the rate of in-migration is relatively high, had not considered locating their businesses in another metro area; 81 percent of the Worcester business owners and 70 percent of those in Colorado Springs said they had not considered locating elsewhere. In Worcester, moreover, women were significantly less likely than men to have considered a location outside of the Worcester metro area (only 14 percent of women vs. 24 percent of men said they had given thought to locating elsewhere; $p=.06$). People may not think of locating a business outside their home metro area because many retain their wage and salary jobs while starting a business; Reynolds and White (1997) found that between 60 and 80 percent of their respondents were still working at a wage and salary job while launching their businesses.

These patterns probably also reflect the residential rootedness of business owners, who are, in fact, more rooted to place than is the general population. As shown in Figure 4, higher proportions of business owners than of the total population had

---

[65] Reynolds and White, The entrepreneurial process *op. cit.*

[66] Because of the large military presence in Colorado Springs, many current residents had lived there previously while serving in the military.

lived in the metro area for at least five years. In addition, women we interviewed in both metro areas had lived longer in place before starting a business than had their male counterparts (see Figure 5). These high levels of rootedness lead most entrepreneurs to be unwilling to move their enterprises elsewhere. Only 11 percent of the entrepreneurs we interviewed in Worcester (19 percent in Colorado Springs) said they would now consider moving their business elsewhere.[67]

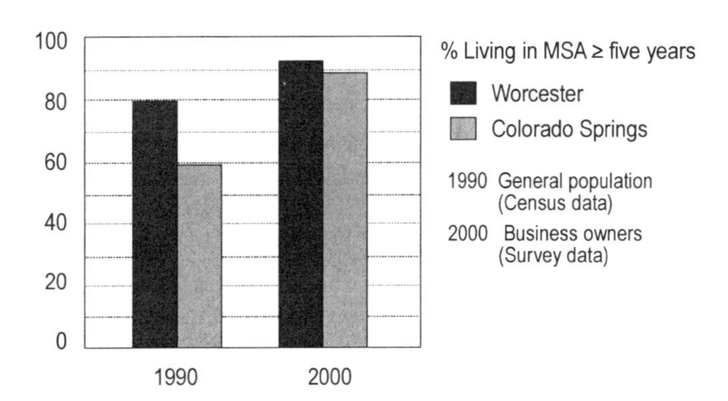

Figure 4 Business owners are more rooted to place than is the general population. Source: U.S. Census of Population 1990; mailed surveys from Worcester, Massachusetts (1998) and Colorado Springs, Colorado (2000)

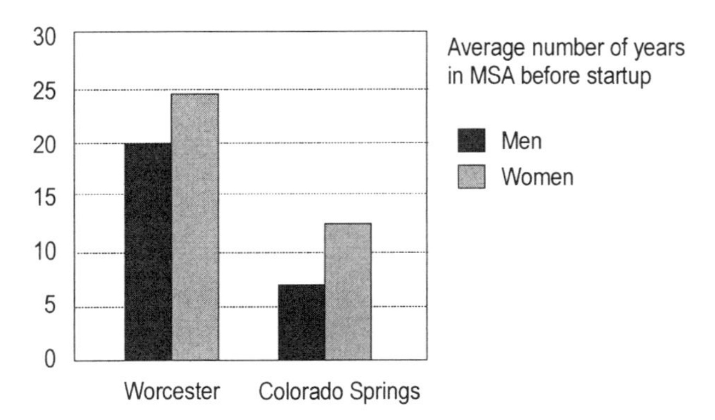

Figure 5 Women entrepreneurs have lived longer in place than their male counterparts before starting a business. Source: Personal interviews, Worcester, Massachusetts (1998) and Colorado Springs, Colorado (2000)

[67] The gender difference in Worcester is significant: 12 percent of women vs. 17 percent of men ($p=.02$).

At a finer scale, many entrepreneurs choose to locate their businesses at home (18 percent of the business owners we interviewed in Worcester and 15 percent in Colorado Springs currently had their businesses in their homes) – or near to home – for reasons having to do more with family or lifestyle than with the nature of the business itself; a home location chosen for financial or family reasons may not actually be a good location for the business.

So the urban area and the part of the urban area where the business is located are dictated by pre-existing residential location; the building is often chosen simply because it was available or offered at the right price, or because the business owner's personal contacts led him or her to choose that location.

In sum, current residential location trumps all other geographies when it comes to locating a fledgling business; nascent entrepreneurs are not scanning distant horizons in search of the best location or even a good one. A careful look at the locational decision making of business owners points to the overwhelming pull of the local, and this pull is especially strong in some places, like Worcester, and in some places, it is especially strong for women.

Impacts of entrepreneurs and entrepreneurship on place

Geographers and economists have been interested in entrepreneurship mainly because of its presumed link to, and driving force behind, local and regional economic development.[68] The type of entrepreneurship that is the topic of such discussions is, however, what I referred to earlier as that which is based in "true innovation" rather than a form of entrepreneurship that may truly affect the quality of life in a place but may not produce an exportable product.

As I described in the earlier section, "what counts as innovation," one way that entrepreneurs change the places in which their businesses are located is by introducing innovations to that place. The shoe inspection firm I described is an example of entrepreneurs starting something that was new to the place, even if it was not new to the world. Many of these "innovative-for-here" businesses, such as the respite care facility I mentioned, provide a product or service that is badly needed and important to community well-being. Indeed, when seen from the perspective of "contribution to place," perhaps the local orientation and small size of many women's businesses – characteristics that traditionally have been viewed as weaknesses – should instead be valued as sources of strength.

All businesses have two direct impacts on the places in which they are located; they provide employment and they pay taxes. But entrepreneurs have additional

---

[68] E.g., Edward J. Malecki, 'Entrepreneurship in regional and local development', *International Regional Science Review* 16 (1994) pp. 119-53.

significant impacts on place beyond employing people and paying taxes. The strategies they use to bring more balance and meaning to their lives are not entirely atomistic or individually oriented; for a great many entrepreneurs, achieving balance includes community involvement.

We found, for example, that business owners volunteer at a rate that is about twice that of the general population: 54 percent of the people we interviewed in Worcester and 46 percent of those in Colorado Springs said they were involved in volunteer activities, almost exclusively within the local community.[69] Moreover, half of those who do volunteer named more than one organization they were involved with (schools, churches, youth organizations, sports, arts, and the like). Despite the time demands of running a business, in both Worcester and Colorado Springs, one-quarter of those who volunteered said that their level of volunteering had increased, and about one-third said that it had remained at the same level, since starting the business. Furthermore, this community involvement is not simply opportunistic on the part of business owners; that is, they are not participating in volunteer activities simply to cultivate the local market: we learned that an entrepreneur's involvement in volunteer activities is not related to the local dependence of the business (measured by percentage of total sales that are made in the local area).

Another way in which almost all businesses contribute to the places in which they are located is through charitable contributions, whether of money or in-kind products or services. At least 90 percent of the owners we interviewed in each place made such contributions from their businesses, and the receiving organizations were almost always local.

Finally, when we asked about their meaning of success, many entrepreneurs in Worcester (but far fewer in Colorado Springs) indicated that a large component of "success" for them had to do with the importance of their business to the community. The kinds of responses that indicated this included things like "we offer an important product or service," "we are important to our customers," or "some people would find it a hardship if we weren't here." Also, this definition of success was more likely to be mentioned by women than by men in both places, but the gender difference is especially strong in Worcester.[70] The key point here for a feminist geographer is that this understanding of success is both gendered and geographic; that is, it varies significantly by both gender and place, suggesting that

---

[69] I do not yet have comparable data on the rate of volunteering in the general population, but political scientist Henry Brady (personal communication) told me that his research suggests that the overall rate is about 25 percent.

[70] In Worcester, 41 percent of women vs. 22 percent of men gave responses that related to the importance of their business to the community; the comparable figures in Colorado Springs were 12 percent of women and 9 percent of men.

theorizing entrepreneurship without geography and gender *and* their intersection impoverishes understandings of entrepreneurial processes.

In sum, it is clear that entrepreneurs are changing the contexts that in part gave rise to their businesses. The relationship between entrepreneurship and place is reciprocal, and probing this dialectical relationship will no doubt prove rewarding.

Conclusions

Entrepreneurship attracts considerable scholarly interest because it is seen as the engine of economic growth and innovation.[71] The relationship of entrepreneurship and self-employment to labor market processes attracts attention, in part because, at least in the U.S., so many people have had periods of self-employment during their working lives; Reynolds and White note that more than 40 percent of people over 60 years of age have been self-employed at some time.[72]

This feminist and geographic consideration of entrepreneurship suggests additional reasons for studying the entrepreneurial process. Perhaps the most important of these are the impacts of business ownership on people's lives and on places. The analysis indicates that in order to understand business decisions and strategies we need to appreciate the many "non-economic" motivations people have for launching a business and we need to recognize how the geography of everyday life is implicated in these motivations, especially via people's desire for greater time-space flexibility in their daily activity patterns.

Our interviews suggest that people are using entrepreneurship as a way to create meaning in their lives, and this observation does not apply only to micro-scale enterprises; it threads through the narratives of owners of large as well as small, and international as well as locally focused, firms. Owning a business provides people with the opportunity to do something they enjoy doing and to gain more autonomy in the workplace. But wanting this greater autonomy does not equate simply to taking the chance to accrue power and money; for many – women and men – it also takes the form of wanting to achieve more balance between and among work, family, and community. It means being able to create a workplace where people will enjoy spending their time and to create communities where they will want to continue to live. I hope that I have helped you to see how geography is completely bound up with the strategies people employ to create more satisfying lives through entrepreneurship and self-employment.

---

[71] Malecki, 'Entrepreneurship in regional and local development' *op. cit.*; Reynolds and White, The entrepreneurial process *op. cit.*

[72] Ibid.

Finally, entrepreneurs' contributions to place seem to be better understood and more fully appreciated by business owners themselves than by scholars of entrepreneurship, who have largely neglected this dimension of entrepreneurship. Clearly, through the very nature of their businesses – many of which, while small, meet important local needs – and through their volunteer activities and their donations to charitable organizations, entrepreneurs are transforming the places in which their businesses are located. If "woman entrepreneur" ceases to be an oxymoron, such that the potential contributions of woman-owned businesses come to be fully appreciated and valued by the gatekeepers (bankers, lawyers, the business elite), perhaps then the contributions of women business owners to place can be fully realized.

Acknowledgments

I thank the Sloan Foundation and the National Science Foundation (grant SBR 9730661) for support of this research and The William and Flora Hewlett Foundation (grant # 2000-5633) for supporting my fellowship at the Center for Advanced Study in the Social and Behavioral Sciences, Stanford, 2001-02, during which I put together the ideas for this colloquium. Many of the ideas in this paper have benefited from discussions with Megan Blake at the University of Sheffield, U.K.; in particular, we are collaborating on a paper about innovation that makes similar arguments to the ones I include here.

KLAUS TSCHIRA FOUNDATION

The Klaus Tschira Foundation gGmbH

Physicist Dr. h.c. Klaus Tschira established the *Klaus Tschira Foundation* in 1995 as a not-for-profit organization designed to support research in informatics, the natural sciences, and mathematics, as well as promotion of public understanding in these sciences. Klaus Tschira's commitment to this objective was honored in 1999 with the „Deutscher Stifterpreis" by the National Association of German Foundations. Klaus Tschira is a co-founder of the SAP AG in Walldorf, one of the world's leading companies in the software industry. After many years on the board of directors, Klaus Tschira is now a member of the company's supervisory board.

The Klaus Tschira Foundation (KTF) mainly provides support for research and student projects in applied informatics, the natural sciences, and mathematics, educational projects at public and private universities and selected projects dedicated to the preservation of historical monuments and the arts. In all its activities, KTF tries to foster public understanding for the sciences, mathematics, and informatics. The resources provided are largely used to fund projects initiated by the Foundation itself. To this end, it commissions research from institutions such as the *European Media Laboratory (EML),* founded by Klaus Tschira in 1997. The central objective of this institute of applied informatics is to develop new information processing systems in which the technology involved does not represent an obstacle in the perception of the user. In addition, the KTF invites applications for project funding, provided that the projects in questions are in line with the central concerns of the Foundation.

The home of the Foundation is the Villa Bosch in Heidelberg, the former residence of Nobel Prize laureate for chemistry Carl Bosch (1874-1940). Carl Bosch, scientist, engineer and businessman, entered BASF in 1899 as a chemist and later became its CEO in 1919. In 1925 he was additionally appointed CEO of the then newly created IG Farbenindustrie AG and in 1935 Bosch became chairman of the supervisory board of this large chemical company. In 1937 Bosch was elected president of the Kaiser Wilhelm Gesellschaft (later Max-Planck-Gesellschaft), the premier scientific society in Germany. In his works, Bosch combined chemical and technological knowledge at its best. Between 1908 and 1913, together with Paul Alwin Mittasch, he surmounted numerous problems in the industrial synthesis of ammonia, based on the process discovered earlier by Fritz Haber (Karlsruhe, Nobel Prize for Chemistry in 1918). The Haber-Bosch-Process, as it is known, quickly became and still is the most important process for the production of ammonia. Bosch's research also influenced high-pressure synthesis of other substances. He was awarded the Nobel Prize for Chemistry in 1931, together with Friedrich Bergius.

In 1922, BASF erected a spacious country mansion and ancillary buildings in Heidelberg-Schlierbach for its CEO Carl Bosch. The villa is situated in a small park on the hillside above the river Neckar and within walking distance from the famous Heidelberg Castle. As a fine example of the style and culture of the 1920's it is considered to be one of the most beautiful buildings in Heidelberg and placed under cultural heritage protection. After the end of World War II the Villa Bosch served as domicile for high ranking military staff of the United States Army. After that, a local enterprise used the villa for several years as its headquarters. In 1967 the Süddeutsche Rundfunk, a broadcasting company, established its Studio Heidelberg here. Klaus Tschira bought the Villa Bosch as a future home for his planned foundations towards the end of 1994 and started to have the villa restored, renovated and modernised. Since mid 1997 the Villa Bosch presents itself in new splendour, combining the historic ambience of the 1920's with the latest of infrastructure and technology and ready for new challenges. The former garage situated 300 m west of the villa now houses the Carl Bosch Museum Heidelberg, founded and managed by Gerda Tschira, which is dedicated to the memory of the Nobel laureate, his life and achievements.

Text: Klaus Tschira Foundation 2003

For further information contact:

Klaus Tschira Foundation gGmbH
Villa Bosch
Schloss-Wolfsbrunnenweg 33
D-69118 Heidelberg, Germany
Tel.: (49) 6221/533-101
Fax: (49) 6221/533-199
beate.spiegel@ktf.villa-bosch.de

Public relations:
Renate Ries
Tel.: (49) 6221/533-214
Fax: (49) 6221/533-198
renate.ries@ktf.villa-bosch.de

http://www.villa-bosch.de/

PHOTOGRAPHIC REPRESENTATIONS

Photographic representations: Hettner-Lecture 2002

Plate 1 & 2 Susan Hanson and Klaus Tschira in the *Alte Aula*.

Plate 3 Musical opening.

Plate 4 Reception in the *Bel Etage*, *Rector's Office*.

Plate 5 Conversation.

Plate 6 Seminar in the *Studio* of the *Villa Bosch*.

Plate 7 Group discussions.

Plate 8 Break.

Plate 9 Outdoor talk.

Plate 10 Debates in the historical gardens of the *Villa Bosch*.

Plate 11 More discussion.

Plate 12 Susan Hanson.

LIST OF PARTICIPANTS

List of participants

The following graduate students and young researchers participated in one or several of the three seminars with Susan Hanson:

ARNTZ, Melanie; Department of Geography, Bonn
BECK, Grit; Department of Geography, Free University, Berlin
BERWING, Stefan; Department of Geography, Heidelberg
DÖRRE, Andrei; Department of Geography, Humboldt-University, Berlin
DORBAND, Jana; Department of Geography, Heidelberg
FELBER, Patricia; Swiss Federal Research Institute WSL, Birmensdorf
FORSTER, Ute; Department of Geography, Heidelberg
FREYTAG, Tim; Department of Geography, Heidelberg
GAMERITH, Werner; Department of Geography, Heidelberg
GOEKE, Pascal; Institute for Migration Research and Intercultural Studies, Osnabrück
HOYLER, Michael; Department of Geography, Heidelberg
JÖNS, Heike; Department of Geography, Heidelberg
KRAMER, Caroline; Department of Geography, Heidelberg
MAGER, Christoph; Department of Geography, Heidelberg
MATTISSEK, Annika; Department of Geography, Heidelberg
MÜLLER, Anke; Department of Geography, Jena
PETHE, Heike; Department of Geography, Humboldt-University, Berlin
RABE, Claudia; Department of Geography, Karlsruhe
SACHS, Klaus; Department of Geography, Heidelberg
SCHIER, Michaela; Department of Geography, University of Technology, Munich
SCHMID, Heiko; Department of Geography, Heidelberg
SCHMITT, Thomas; Dillingen/Saar
STREIT, Anne von; Department of Geography, University of Technology, Munich
WEIST, Thorsten; Department of Geography, Humboldt-University, Berlin
WINTZER, Jeannine; Department of Geography, Jena
WOLKERSDORFER, Günter; Department of Geography, Münster

Plate 13 Some participants of the Hettner-Lecture 2002.

HETTNER-LECTURES

1 *Explorations in critical human geography* DEREK GREGORY 1997

2 *Power-geometries and the politics of space-time* DOREEN MASSEY 1998

3 *Struggles over geography: violence, freedom and development at the millennium* MICHAEL WATTS 1999

4 *Reinventing geopolitics: geographies of modern statehood* JOHN A. AGNEW 2000

5 *Science, space and hermeneutics* DAVID N. LIVINGSTONE 2001

6 *Geography, gender, and the workaday world* SUSAN HANSON 2002

Please order from: *Franz Steiner Verlag GmbH / www.steiner-verlag.de*
Distribution by Brockhaus / Commission, Kreidlerstraße 9, D-70806 Kornwestheim
E-Mail: bestell@brocom.de Tel. 0049 (0)7154 1327-0 Fax 0049 (0)7154 1327-13